SpringerBriefs in Molecular Science

Biometals

Series Editor

Larry L. Barton

For further volumes:
http://www.springer.com/series/10046

Anil K. Suresh

Metallic Nanocrystallites and their Interaction with Microbial Systems

 Springer

Anil K. Suresh
Department of Molecular Medicine
Beckman Research Institute
City of Hope
Flower Avenue 1710
Duarte, CA 91010
USA

ISSN 2191-5407   e-ISSN 2191-5415
ISBN 978-94-007-4230-7   e-ISBN 978-94-007-4231-4
DOI 10.1007/978-94-007-4231-4
Springer Dordrecht Heidelberg New York London

Library of Congress Control Number: 2012933319

Printed on acid-free paper

Springer is part of Springer Science+Business Media (www.springer.com)

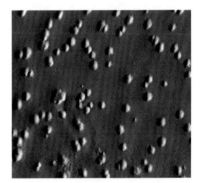

Give me a place to stand, and I will move the Earth

—Archimedes

I dedicate this book to my beloved Ph.D. mentor, Dr. M. I. Khan, my parents, my wife, my brothers, and my lovely little angel daughter Akanksha

Preface

Nanoscience and nanotechnology are a fast growing and dynamic areas, which include novel class of materials that are being developed for various applications. Nanotechnology has immense potential in almost every field of science and technology, primarily due to their size and/or shape dependent intrinsic physico-chemical, optoelectronic, catalytic and biological properties and greater surface area. As some may know, modern nanotechnology originated in the year 1959 after an oral presentation given by Dr. Richard Feynman, *"There's plenty of room at the bottom."* The impetus for modern nanotechnology was provided by inter-facing nanoscience with biology, medicine, electronics and advanced analytical tools. Researchers and industrialists believe that one day nanotechnology will likely impact the perspective of things being looked at, and will drastically rev-olutionize the industries and pharmaceutical companies with great emphasis on human health, environment safety and sustainability. Nanotechnology has already begun to improve many facets of science and technology, and researchers are revisiting several useful aspects with a nanoperspective to understand how similar things could work at the nanoscale. This phenomenon is likely to revolutionize pharmaceutical sciences, and many drugs are being reconsidered for possible deliveries using smart multifunctional nanomaterials.

This book emphasizes two distinct but interrelated and novel aspects with respect to nanoparticles: ecologically benign and cost effective production of nanoparticles, and issues related to safety concerns of nanoparticles on the biotic environment. Nanoparticles in distinct forms are extensively used in various consumer products as additives and therefore are required in huge quantities. In that respect it has become highly imperative to be able to produce nanoparticles at the mega scale, using both ecologically friendly and economic procedures. Also, as the nanoparticles are getting implemented more and more widely, they are released into the environment in one form or another, following it a host for new potential health issues. To prevent this risk, one must look at the proper devel-opment and use of these nanomaterials and the fate, transport and impacts of such engineered nanostructures on the environment must be addressed. Interactions between the nanoparticles and microorganisms in the environment are unavoidable, but the pandemic consequences of such interactions are beginning to be investigated. This book will also illustrate how naturally occurring microorganisms and manmade nanoparticles interact, and the consequences of such interaction, using suitable examples from our studies published in several peer reviewed International Journals. Because of its uniqueness in content and scope, I am positive that this volume will be helpful not only to the scientific and industrial community but it will also attract the attention of students and researchers in different areas of sciences such as microbiology, biotechnology, nanotechnology, toxicology, materials science, biomedical engineering, cell and molecular biology etc. The several objectives of this brief are to introduce nanobiotechnology along with the fast emerging "green biosynthesis" for their manufacture, and to let the readers aware on the potential interactions of engineered nanoparticles with microorganisms. Impacts of noble metal and metal oxide nanoparticles such as gold, silver and cerium oxide on the growth and viability of several Gram-negative and Gram-positive bacteria will be presented. Differences in the interactions using different forms of nanomaterials, nanoparticles synthesis methodology, surface coatings, and the various analytical assays used to determine the bactericidal activity will be described. Mechanistic insights on the relationship between the bacterial growth inhibition, reactive oxygen species generation and up and/or down regulation of transcriptional stress responsive genes will also be discussed. Finally, how we made use of the emerging advance imaging techniques such as transmission electron and atomic force microscopes that will shed impacts towards a better understanding on the overall microbial–nanoparticle interactions will be discussed.

Overall, the book contains five chapters. Chapter 1 includes the basic and general introduction to nanoscience and nanotechnology, properties of nanoparticles, synthesis methodologies employed to produce various nanoparticles, physical characterizations of nanoparticles, and the applications of nanoparticles, with emphasis on biological and medicinal applications. Chapter 2 details the microbial based biofabrication of nanoparticles, mechanism involved behind biofabrication, and the advantages of biosynthesis over the existing conventional chemical and physical routes of synthesis. Moreover, the reliability of biosynthesis technique with detailed description on the biosynthesis with suitable example, thorough physical characterizations of the synthesized particles so as to assess their morphology, crystallinity, surface characteristics based on advanced analytical tools will be presented. Chapter 3 discusses the bactericidal properties of engineered metal nanoparticles and analysis the comparative toxicity assessments of engineered silver nanoparticles on bacteria and discusses the toxicity assessments of nanoparticles, deemed reasons for nanoparticles being considered toxic and the necessity to address the potential toxicity of nanoparticles. It then proceeds to describes the external factors that might govern nanoparticles mediated toxicity and the proposed mechanisms behind the toxicity. Additionally, it discusses the details of the various techniques used to evaluate bactericidal activity, their advantages and limitations along with the influence of size, shape, surface coatings of nanoparticles on the toxicity and the mechanistic of bacteria-nanoparticle interactions. Chapter 4 focuses on the biocompatibility and inertness of gold nanocrystallites and analyses the inert nature of gold nanoparticles along with its biosynthesis and physical characterizations. Last but not least, Chap. 5 examines the antibacterial properties of engineered metal oxide nanocrystallites and the stress mechanism involved. As an example, it describes our work on the effects of engineered cerium oxide nanoparticles on the growth and viability of several Gram-negative and Gram-positive bacterial strains. It then discusses the relation between the growth inhibition, reactive oxygen species generation and up and or down regulation of transcriptional stress genome. Finally, it analyses the use of advanced analytical tools like the transmission electron microscopy to evaluate the bacterial response mechanisms.

I am pleased that I have been invited to write this book published by Dr. Sonia Ojo, senior publishing editor at Springer within the Springer Briefs in Biometals series by Prof. Larry Barton. To all I wish a happy reading!

Duarte, 12 December 2011   Anil K. Suresh

Acknowledgments

Much to my extreme delight, I would like to evince my gratitude and indebtedness to my beloved Ph.D. mentor *late. Dr. M. I. Khan*, a truly remarkable scientist who introduced me to this fascinating realm of Nanobioscience and Nanobiotechnology. His invaluable guidance, constant inspiration, and unending support have always been contagious and motivational throughout my Ph.D. pursuit. His scientific temperament, innovative approach, dedication towards his profession and his down to earth nature has inspired me highly. Although this eulogy does not give him justice, I preserve an everlasting gratitude for him.

I heartedly thank *Prof. Jay Nadeau* at McGill University, Canada and *Prof. Yves-Alain Peter* at Ecole Polytechnique in Montreal, Canada, for their mentorship during my first postdoctoral training. Special thanks to them for giving me the opportunity to let me explore my research expertise on nanomaterials on further implementation in cell imaging and drug delivery systems. They often used to organize family get-together and fun-filled extracurricular activities (kayaking, canoeing, skiing, rock climbing, and ropes courses). It was a very friendly environment and I learned a lot from them.

I wish to express my sincere gratitude and heartfelt thanks to *Dr. Mitchel Doktycz* and *Dr. Dale Pelletier* at Oak Ridge National Laboratory, USA. With them I pursued my second postdoctoral training. I am grateful for their mentorship, motivation, subtle guidance, fruitful discussions, never ending support, and constant help. The trust and freedom they gave me to implement my own research ideas have been crucial to achieve this feat. Working with them has always made me feel relaxed and has enabled me to progress in a lively and cool environment. I will never been able to thank them enough. They are like godfathers to me.

I would also like to use this opportunity to express my sincere thanks to *Prof. Jacob Berlin*, not only for giving me the opportunity to work as a Staff Scientist in his Laboratory at the Department of Molecular Medicine, Beckman Research Institute at City of Hope, USA but also for introducing me to my dream area of research: *Cancer Therapeutics and Clinical Medicine*.

I sincerely acknowledge and I am very much thankful to my research collaborators and colleagues *Dr. Tommy Phelps, Dr. Wei Wang, Dr. Aloke Kumar,*

Dr. Ji-Won Moon, Dr. Baohua Gu and Prof. David Allison, at the Oak Ridge National Laboratory for their valuable help in generating data and editing manuscripts, for our fruitful discussions, and for showing me their constant support, inspiration and motivation throughout my stay at ORNL.

I would also like to thank all my lab mates and colleagues I worked with throughout my research period at National Chemical Laboratory in India; at the McGill University in Canada; at the Oak Ridge National Laboratory in the USA and presently here at City of Hope in the USA, for their constant support and help whenever it was required.

It gives me immense pleasure to thank *Amma, Daddy*, and my *lovely brothers Sunil* and *Vinil*, for their love, unfailing support, tremendous patience, trust and encouragement shown in their own special way during my long period of studies. They have been a constant source of strength and inspiration for me. My due thanks to them for their love, support and faith in me. Also thanks my *sister-in-laws*, who recently joined our family, for their support.

I would also like to thank my *wife, Arundhati*, for her care, understanding, love and for everything she does for me.

I am ever grateful to the *Almighty God*, the Creator and the Guardian, and to whom I owe my very existence; because of his blessings, wisdom and perseverance that he has been bestowing upon me at all times. I bow to the divine strength and hope that his blessings will dwell throughout my life.

Last but not least, thanks to my *daughter Akanksha*, lovely little angel, whose cute smiles and funny acts soothed the pain I experienced while achieving every feat of my life.

Contents

Chapter 1
Introduction to Nanocrystallites, Properties, Synthesis, Characterizations, and Potential Applications

Abstract Nanoscience and nanotechnology include a novel class of engineered nanomaterials that is gaining significant recognition to pursuit in the existing and emerging biological, medical, and engineering applications. This chapter will provide a brief overview of nanotechnology with an emphasis on nanoparticle properties, novel methodologies for nanomaterial synthesis, and the various techniques used to assess nanoparticle characteristics such as size, shape, surface properties, purity, and crystallinity. Finally a broad outline of the numerous applications of nanoparticles will be described, with primary focus on biotechnological and biomedical applications.

Keywords Biomedical · Nanoparticles · Nanoparticle characterization · Nanoparticle synthesis

1.1 Nanoscience and Nanotechnology: An Overview

Nanoscience and nanotechnology involve controlled miniaturization of materials at the nanometer scale where its properties are significantly different from that of their respective bulk counterparts. The term is also referred to the design, production, and implementation of nanostructured devices and systems at the nanoscale. In other words, it is an area that focuses on the production of structures and materials with control over their properties during miniaturization and use of such properties for the development of better functional materials and smart devices [1]. The term nanoparticle is used to describe a wide variety of materials with submicron size distribution. As per British Standard Institute definitions "Nanoparticles are the particles with one or more dimensions at the nanoscale (<100 nm)". At this length scale materials start to behave completely different and the properties of matter differ from atomic and molecular properties that are governed by

A. K. Suresh, *Metallic Nanocrystallites and their Interaction with Microbial Systems*,
SpringerBriefs in Biometals, DOI: 10.1007/978-94-007-4231-4_1,
© The Author(s) 2012

the laws of classical quantum mechanics and/or physics. Therefore the dimensions between 1 and 100 nm might be considered as an intermediate state between atomic or molecular state and the bulk state, where materials exhibit unusual and unexpected new properties that cannot be defined by classical laws of physics. And it is these unique properties that attract wide attention of researchers from almost every discipline of science.

Research in nanotechnology is as diverse as other fields of science such as physics, chemistry, material science, microbiology, biochemistry, and molecular biology. Hence, nanoscience in combination with biotechnology and biomedical engineering is an emerging area related to the development of novel nanostructured materials for imaging, diagnosis, gene sequencing, and drug delivery applications [2]. Thus, nanotechnology can enable one to learn more about the detailed functioning of individual cells and neurons, which in turn will be useful in re-engineering and designing of better living systems. Another interesting aspect of miniaturization is that many factors can influence the physicochemical, optoelectronic, and biological properties of nanomaterials. These factors can in turn strongly modulate their properties viz. size, shape, surface composition, dielectric environment, and interparticle interactions [2]. The factors responsible for such remarkable variations in their properties are their small dimensions and large surface area, and availability of greater surface area facilitates better catalytic efficiency per unit volume of a catalyst [3]. Historical evidence suggests that nanomaterials have been in biomedical use since the ancient times. Gold nanocrystallites in the form of colloidal gold were used as drugs by several Asians in the early 2500 BC. Reports also suggest that colloidal gold namely "Swarna Bhasma" and "Makaradhwaja" are still in use in Indian traditional medicine called the "Ayurveda" that dates back to first millennium BC. During the sixteenth century in Europe, colloidal gold named as "Aurum Potabile (drinkable gold)" was believed to have curable properties to several diseases [4]. However, in the past 12 years, nanomaterials in conjunction with biotechnology and biomedical engineering have led to enormous applications that range from drug and gene delivery to fluorescent labeling, cell and molecular imaging, pathogen detection, pathogen destruction, tumor destruction, tissue engineering, catalysis, solar systems, biosensors, and optical devices [2]. These studies have all been successful due to the tireless efforts by various researchers' across the globe that led to the generation of novel smart materials for intended applications. Nanoparticles with distinct characteristics are required for different types of implications, for example, magnetic resonance imaging (MRI)-based contrast agents, positron emission tomography (PET), and Optical Coherent Tomography-based detection, surface plasmon resonance (SPR)-based optical imaging, and fluorescence-based biolabeling and detection. Although every methodology has its own unique advantages and drawbacks, not all nanoparticles can be used for every application, as the properties of nanomaterials also vary with the types or forms of nanomaterial. For example, CT, MRI, and PET are considered ideal for in vivo imaging experiments, whereas optical- and fluorescence-based imaging are mostly applied for in vitro imaging studies. Finally, it is a well-known fact that "living cells" are the best

Table 1.1 An overview on the various applications of metal, metal oxide, and metal sulfide nanoparticles

Nanoparticle type	Properties owned	Applications
Au spheres	Optical, electronic, catalytic	Drug/gene delivery, imaging, molecule tracking, biosensors
Au rods	Optical, radiative, and non-radiative	Photodynamic therapy, imaging drug delivery
CdS, CdSe-ZnS	Optical, fluorescence, luminescence and electronic	Imaging, diagnostics, molecule tracking, biolabeling, biosensors
Ag, ZnO, TiO_2	Bactericidal, catalytic, piezoelectric, ceramic	Antimicrobial agents, sensors, in pigments
CeO_2	Ceramic, catalytic	Bone transplant, fuel exhaust, catalysis, protecting cells
Fe_3O_4	Contrast	Drug/gene delivery, imaging, MRI contrast agent, positron emission tomography
ZnO	Catalytic, optical, wide band gap	Catalysis, drug and gene delivery, imaging, antibacterial agents
Cu	Catalytic, conductive, tensile	Catalysis, electronics, thin films
Pt, Pd	Catalytic, thermo-electric, plasmonic	Sensors, electrode positing, fuel cells

examples of machines that operate at the nanoscale performing numerous functions, ranging from generation of energy to extraction of targeted materials with very high efficiency. Hence, the scientific interest in this interface of biology and technology is based on the perception that nanobiotechnology can potentially lead to the development of novel interesting and useful functional nanosystems.

1.2 Properties of Nanomaterials

Miniaturization is known to have pronounced effect on the materials physical properties, which are different from that of the corresponding bulk materials. Some of the physical properties exhibited by nanomaterials are due to (1) large surface area or surface energy, (2) spatial confinement, and (3) reduced imperfections. An illustration on the unique properties owned by some of the novel metal and metal oxide nanoparticles are given in Table 1.1, and are briefly described below:

1.2.1 Optical Properties

The visible property of metallic nanoparticles is their colored colloidal suspensions. Mie explained the red color of gold nanoparticles on the basis of Maxwell's equation for an electromagnetic light wave interacting with small metallic spheres [2]. The color exhibited by metallic nanoparticles is due to the coherent excitation of the "free" electrons within the conduction band, leading to an in-phase oscillation known as surface plasmons, the coherent charge density oscillations. The excitation of surface plasmons by an electromagnetic field at an incident wavelength, where resonance occurs, results in the appearance of intense surface plasmon resonance (SPR) bands and an enhancement of local electromagnetic field. Such quantum size effects are best studied with metal and semiconductor nanoparticles, where the energy level spacing for a spherical particle is predicted to be inversely proportional to R^2. Thus, with decrease in the size the effective band gap increases, leading to relevant absorption and emission spectra blue shifts. Hence, the difference in the color of metallic nanoparticles can be correlated to their size and surface plasmon resonance and has been exploited for various applications; for example, the change in color of gold nanoparticles from ruby red to blue due to aggregation has been exploited for the development of highly sensitive colorimetric DNA analysis technique.

Unique optical property of nanomaterials can also be due to quantum size effect, which arises primarily due to the confinement of electrons within particles of dimension smaller than the bulk electron delocalization length. This effect is more pronounced for semiconductor nanoparticles, where the band gap increases with a decrease in size. The same quantum size effect is also shown by metal nanoparticles, when the particle size is >2 nm.

1.2.2 Fluorescence Properties

Semiconductor nanoparticles also known as quantum dots are the single nanocrystallites of less than 10 nm, which process size-dependent absorption and emission shift. Absorption of a photon with energy above the semiconductor band gap energy results in the creation of an electron–hole pair or exciton. The absorption has an increased probability at higher energies, i.e., shorter wavelengths, and results in a broadband absorption spectrum, in marked contrast to standard fluorophores. For nanocrystals smaller than the Bohr exciton radius (a few nanometers), energy levels are quantized, with values directly related to the quantum dot size, an effect called quantum confinement, hence the name "quantum dots". The radiative recombination of an exciton characterized by a long lifetime of 910 ns leads to the emission of a photon in a narrow, symmetric energy band, different from the red-tailed emission spectra and short lifetimes of most conventional fluorophores. The long fluorescence lifetime of quantum dots enables the use of time-gated detection to separate their signal from that of shorter lived species, such as background auto-fluorescence encountered in cells therefore making them promising detection probes.

1.2.3 Magnetic Properties

Magnetic properties of nanostructured materials are distinctly different from that of bulk materials. A magnetic property of a single isolated particle is strongly influenced by the particle size. Decrease in particle size to a certain size range, where the particles consist of only a single magnetic domain, the coercitivity, and remanence are increased drastically. Further decrease in the particle size leads to a decrease of remanence and coercitivity to zero. This transformed paramagnetism behaves differently from that of conventional paramagnetism, and is referred to as "superparamagnetism" and is observed when the thermal energy of the particle kT is larger than the energy of magnetic anisotropy KV. Ferromagnetic particles become unstable when the particle size is reduced below a certain limit and increase in surface energy provides sufficient energy for domains to spontaneously switch polarization directions to become paramagnetic. In short, ferromagnetism of bulk material disappears and gets transformed to superparamagnetism at the nanoscale due to high surface energy.

1.2.4 Mechanical Properties

Materials are often subject to external mechanical forces (elasticity, ductility, shear, toughness, tensile strength) that depend on how materials deform (elongate, compress, and/or twist) or break as a function of applied load, time, temperature, and other conditions. These forces in turn depend on several factors such as size and shape of the material, how the material is held, and the way of performing mechanical shear. The mechanical properties of nanomaterials are known to increase with decrease in size. The enhanced mechanical strength is due to its high internal perfection. Generally, imperfections such as dislocations, micro-twins, impurities, etc. in crystals are highly energetic and needs to be eliminated from perfect crystal structures. Smaller the cross-section of nanowires less is the probability of any imperfections and nanoscale dimension permits the elimination of such imperfections, thereby with better mechanical properties.

1.2.5 Thermal Properties

Metal and semiconductor nanoparticles have lower melting point or phase transition temperatures when compared to their bulk counterparts. The lowering of the melting point is observed when the particle size is >100 nm and is attributed to increase in the surface energy size reduction. The decrease in the phase transition temperature can be correlated to the changes in the ratio of surface to volume energy as a function of size.

1.2.6 Catalytic Properties

Catalysis is a central concept of chemistry, playing a prominent role in various biological, biomedical, materials science, and industrial processes. Catalysis is defined as a change in the rate of a chemical reaction due to the participation of a substance called a catalyst. Unlike other reagents that participate in the chemical reaction, a catalyst is not consumed by the reaction itself and therefore may participate in multiple chemical transformations. The advent of nanosciences has now clearly promoted the bottom-up strategy over the top-down, making this traditional frontier obsolete. Thus, the molecular approach is currently most beneficial for defining selective and efficient heterogeneous catalysts that can be removed from reaction medium and re-used. Nanocrystals of size 1 nm to only a few nanometers present the best catalytic efficiency, yet their support is most important for the synergistic activation of substrates. Though several nanoparticles such as Au, Cu, Pd, Pt, Ce, Ru, and Ir are shown to have catalytic properties, gold is considered as the most efficient nanoparticle catalyst, due to its ability to catalyze a variety of aerobic (cost-effective) as well as oxidation reactions.

1.3 Synthesis of Nanocrystallites

Widespread applications of engineered nanoparticles have led to the invention of several approaches for the production of advanced functional nanomaterials. The fabrication of engineered nanomaterials and smart devices have all been achieved utilizing solids, liquids, precursors, biomaterials, microorganisms, and plant extracts that encompass a wide varied set of experiments and techniques. Though the synthesis is predominantly using chemical methods, attempts are also being made to utilize physical methods and naturally existing biological systems viz. microorganisms and biomolecules, incorporating diverse experiments and techniques which is beyond the scope of this brief chapter. Nanoparticles with uniform size and shape distributions were initially introduced by the groups of Turkovich, Frens, Stober, Iijima, Bawendi, and several others to produce nanostructures of gold, silica, carbon, cadmium, etc. However, majority of the currently used synthesis methods can be classified into two main approaches; "bottom-up" and the "top-down" [2]. Bottom-up approach also known as the self-assembly includes the chemical and the biological synthesis; it is an approach where a structure is built either atom-by-atom or molecule-by-molecule or cluster-by-cluster. In this methodology, the building blocks are first fabricated and then subsequently assembled into the final material, whereas in the top-down approach the particles are synthesized by either physical or chemical or mechanical reduction of the starting material. This approach usually leads to nanoparticles with crystal defects and with surface imperfections. Such crystal defects are known to have greater effects on their properties due to high aspect ratio, whereas, using the "bottom-up" method nanoparticles with desired size and shape distributions and particles with no crystal defects can be produced [2].

1.3.1 Synthesis Using Chemical and Physical Methods

Over the past decades various physical and chemical methods have been developed for the production of a vast variety of metal, metal oxide, and metal sulfide nanocrystallites. Currently, different forms of nanoparticles are synthesized using a number of techniques involving either reduction and or oxidation, and precursor mediated precipitation. Turkevich et al. for the first time showed the synthesis of triangular gold nanocrystallites in templateless and seedless medium, where the reduction of auric chloride ions was performed using sodium citrate [5]. A similar methodology was employed by Caswell et al. who demonstrated the synthesis of silver nanowires by using sodium citrate as a reducing agent, in the presence of NaOH at 100°C [6]. Suito and Udeya showed that plate-like structures of gold nanoparticles could be obtained by reducing gold ions with salicylic acid. In this method, the formation of plate-like structures was correlated to spiral growth mechanism along with aggregation and recrystallization of the metal ions [2]. Similarly, silver nanoprisms were synthesized by boiling silver nitrate ($AgNO_3$) in N,N-dimethylformamide in the presence of poly-(vinylpyrrolidone). Likewise, implementing similar methodology the synthesis of nanowires of silver and platinum and nanocubes and nanotriangles of silver were also achieved [7, 8].

Some of the physical methods employed includes the sol–gel techniques, solvothermal synthesis, inert gas condensation, laser ablation, photoirradiation, aerosol techniques, lithography, radiolysis, ultrasonication, physical vaporization, spray pyrolysis, attrition, and solvated metal atom dispersion [2]. Inert gas condensation is usually performed in closed chambers and used for the creation of very small materials at the atomic levels. Due to the interatomic collision with the gas atoms within the chamber, the evaporated metal atoms tend to lose their kinetic energy and therefore condense into small crystallites [9]. In laser ablation methodology, pulsed laser irradiation is used on the targeted metal either using liquid and/or gas atmosphere to synthesize nanoparticles [10], whereas in pyrolysis, the precursor suspension is atomized into a series of very small droplets called the "reactors", which were then introduced into a hot wall region by a carrier gas under high atmospheric pressure. The solvent in the droplets evaporate inside the furnace, thereby causing precipitation, thermal decomposition followed by interparticle interaction of the solute to produce nanoparticles [11]. Jin et al. described the transformation of spherical-shaped silver nanoparticles to triangular shapes by photo irradiation [12]. A thermally induced transformation was also employed for the conversion of spherical gold clusters into nanocubes [12]. In an interesting approach, Roorda et al. synthesized gold nanorods by ion beam irradiation of spherical gold–silica core–shell nanoparticles [13]. The ion beam irradiation was shown to deform the silica shell leading to the formation of the core spherical gold nanoparticles.

The control of size, shape, stability and the assembly of nanoparticles can be achieved by incorporating different additives in the form of capping agents, precursors, solvents, and templates. Transformation of nanoparticles with respect to size and shape distributions primarily depend on the capping agent used as stabilizer and range from simple ions to detergents, biomolecules, organics, and biopolymers. Though water is the most commonly used solvent, reports also exist on the use of various organic solvents [2]. Apart from ionic liquids and super-critical fluids, micelles and organic polymers are also employed for the synthesis of nanomaterials. Some of the common biopolymers used are the deoxyribonucleic acid, nucleotides, proteins and peptides, and sugars. Furthermore mesoporous materials and preformed nanoparticles have also been utilized as seeds for the controlled assembly and formation of nanocrystallites [2].

Currently, increased attention is being directed toward controlling the shape of the nanoparticles rather than the size. Nanoparticles of various anisotropic shapes viz. rods and wires, tubes, dumbbells, cubes, hexagons, tetrahedrals, decahedrons, multipods, star shaped, disks, triangles, and dendritic shaped are produced employing various synthesis strategies that include: synthesis in micellar/surfactant suspensions, use of soft and rigid templates, controlling the growth using physical confinements, use of nanosphere lithography, vacuum vapor deposition, direct synthesis in solution in presence or absence of additives, and morphological transformation of the preformed nanoparticles by thermal and photo or ion irradiation processes [2]. Some of the most commonly used shapes of nanoparticles for various applications are shown in Fig. 1.1. Rod-shaped nanoparticles, with unique near infrared (NIR) absorption and thermal properties, have been synthesized in the presence of various surfactants by electrochemical, photochemical, and seed-mediated methodologies. Wang et al. synthesized gold nanorods using electrochemical method in a micellar solution of hexadecyltrimethylammonium bromide (CTAB), and using gold metal plate as an anode and platinum plate as cathode [14]. By this method, the bulk gold metal ions from anode were converted into gold nanoparticles at the interfacial region between cathode surface and the electrolyte solution. Esumi et al. showed the synthesis of gold nanorods by photochemical irradiation in a micellar solution of hexadecyltrimethylammonium chloride (HTAC) [15]. The authors mentioned that the formation of elongated gold nanorods instead of spheres occur only when enough concentration of HTAC is used to induce the formation of rod-like micelles in solution. Further Kim et al. suggested that the overall yield of the gold nanorods could be improved by the addition of varying amounts of $AgNO_3$ to the reaction mixture [16].

Though, similar observations were also made by Murphy et al. their synthesis was based on the reduction of $AuCl_4$ ions in presence of CTAB as a surfactant instead of HTAC, and using preformed spherical nanoparticles as seeds [17]. Moreover, ascorbic acid was used as a reducing agent as it reduces $AuCl_4$ ions to stable Au^0 in the presence of CTAB only after the addition of gold seeds, thereby preventing fresh nucleation. According to the authors, slight modification in the experimental conditions might lead to the production of nanomaterials of diverse morphologies such as blocks, cubes, and tetrapods. The formation of rod-shaped nanoparticles was explained on the basis of either templating action of the micelles or preferential binding of CTAB molecules on the faces of fcc {110} gold [18]. Busbee et al. noted that gold nanorods of larger size distributions are formed only

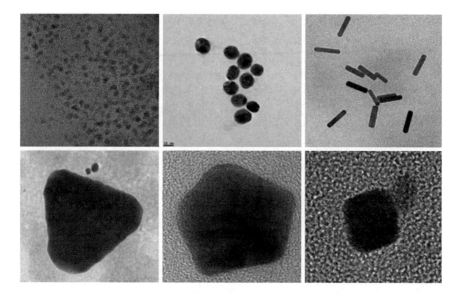

Fig. 1.1 Transmission electron micrograph images of the commonly used shapes of nanoparticles, obtained using different synthesis methodologies

when small gold seeds are used [19]. Murphy et al. observed that the use of Br$^-$ ions as counter ions are essential for the formation of nanorods, whereas Cl$^-$ and I$^-$ ions were found to be noneffective [17]. Pileni synthesized copper nanoparticles of different shapes viz. cubes, rods ,and triangles using copper (II) bis (2-ethylhexyl) sulfosuccinate Cu-(AOT)2-isooctane-water system and salts of different anions as the reaction medium [20]. Their results showed that, though the AOT micelle structure did not change much in presence of different salts, a drastic change in the nanoparticle morphology was observed. A method based on seeded growth in an aqueous suspension containing the capping agent CTAB, gold ions, ascorbic acid, and NaOH was demonstrated for the synthesis of gold nanotriangles [21]. Zheng et al. synthesized silver nanowires and dendrites in a mixture of CTAB and sodium dodecyl sulfate by the reduction of AgNO$_3$ with ascorbic acid. According to the authors by changing the experimental conditions viz. concentration of ascorbic acid and the presence or absence of NaCl led to the formation of one-dimensional nanowires and dendrites [22]. Nanorods of various metals have also been synthesized within the pores of rigid materials. For example, polycarbonates and porous alumina membranes have been used for the synthesis of metal nanotubes and nanowires either by chemical reduction or by electrochemical deposition [2]. The physical constraints of liquid–liquid and air–water interface have also been exploited for the formation of gold tapes, plates, and fractal structures [23]. Though all these synthesis methods have their own advantages, drawbacks do exist that are discussed in the following section. To overcome this, an alternative approach which exploits the synthesis of nanomaterials using the naturally existing biological materials [2] is used.

1.3.2 Synthesis Using Biological Methods

As discussed above, considerable efforts have been devoted by various researchers for the generation of different types of nanomaterials. Most of the above-discussed methods, however, require expensive equipment (e.g. laser), are cumbersome, and ecologically unfriendly, involve the use of toxic solvents and surfactants, combustible precursors, and are accomplished under oxygen and/or a water-free atmosphere requiring high temperatures and pressures. Moreover the resulting nanoparticles can become unstable upon interaction with biomolecules. To overcome these aforementioned issues of physical and chemical methodologies, an alternate synthesis route involves the utilization of naturally existing materials for the reduction of metal ions into stable nanocrystallites. Details on the various biosynthesis techniques used to fabricate diverse nanoparticles are described later in Chap. 2.

1.4 Characterization of Nanomaterials

Progress in nanotechnology has also led to the discovery of novel advanced analytical tools and techniques. Some of the most widely used techniques to assess and evaluate nanoparticle characteristics such as the purity, size, shape, surface properties, and crystallinity are briefly illustrated below.

1.4.1 X-ray Diffraction

Physicists Sir W. H. Bragg and his son Sir W. L. Bragg developed a relationship to explain the cleavage faces of crystals that appear to reflect X-ray beams at certain angles of incidence (θ). This observation is an example of X-ray wave interference, commonly known as X-ray diffraction (XRD), and was a direct evidence for the periodic atomic structures of crystals. Since then XRD is used to determine the crystal structures of various solids, including lattice constants and geometry, orientation of single crystals, and crystal defects. Bragg's equation obtained by measuring the diffraction pattern of the crystals correlates the distance between two atomic layers or *hkl* planes (d) and the angle of diffraction (2θ) in a crystal as $n\lambda = 2d\sin\theta$, where, λ = wavelength of X-rays, n = an integer known as the order of reflection (*h*, *k* and *l*), which represent Miller indices of the respective planes. Moreover, the average size of the nanoparticles can be estimated using the Debye–Scherrer equation: $D = k\lambda/\beta\cos\theta$, where D = thickness of the nanocrystal, k is a constant, λ = wavelength of X-rays, β = width at half maxima of (111) reflection at Bragg's angle 2θ [2].

1.4.2 Scanning Electron Microscopy

Scanning electron microscopy (SEM) is a microscope that uses electrons instead of light to form an image and is one of the most widely used technique for the characterization of nanostructures. The SEM has many advantages over traditional microscopes; has a large depth of field and higher resolution, which allows larger specimen to be in focus at one time simultaneously, closely spaced specimens can be magnified at much higher levels. As SEM uses electromagnets instead of lenses there is much more control in the degree of magnification. Unlike optical microscopy, this technique not only provides topographical information but also can analyze the chemical composition near the surface. The interaction between the electron beam and the sample gives different types of signals providing detailed information about the surface characteristics, structure, and morphology. When an electron from the beam encounters a nucleus in the sample, the resultant columbic attraction will lead to deflection in the electron's path known as Rutherford elastic scattering. A fraction of these electrons are then backscattered resulting in re-emergence from the incident surface. Since the scattering angle depends on the atomic number of the nucleus, the primary electrons arriving at a given detector position produce image yielding topological and compositional data. The high-energy incident electrons can also interact with loosely bound conduction band electrons in the sample. However, the amount of energy given to these secondary electrons as a result of such interactions is small with a very limited range. The secondary electrons produced within a very short distance from the surface escape from the sample giving high-resolution topographical images.

1.4.3 Transmission Electron Microscopy

Transmission electron microscope (TEM) operates on the same basic principles as the light microscope but uses electrons instead of light. As TEM uses electrons as "light source" their much lower wavelength makes it possible to get a resolution that is thousand times better than that of a light microscope. TEM is typically used for high-resolution imaging of thin films of solid samples for structural and compositional analysis. The technique involves: (1) irradiation of a thin film by a high-energy electron beam, which is then diffracted by lattices of the crystalline material and propagates in different directions, (2) imaging and angular distribution analysis of the forward-scattered electrons, unlike SEM where backscattered electrons are detected, and (3) energy analysis of the emitted X-rays. The topographic information obtained by TEM, in the vicinity of atomic resolution, can be utilized for structural characterization and identification of various forms of nanomaterials, viz., hexagonal, cubic, or lamellar. One of the limitations of TEM is that the electron scattering information on TEM originates from a three-dimensional image which is then projected onto a two-dimensional detector.

Therefore, structural information along the electron beam direction is superimposed at the image plane.

Selected-area electron diffraction (SAED) offers a unique method to determine the crystal structure of nanomaterials such as nanocrystalline spheres and rods. In this method, the condenser lens is defocused to produce parallel illumination on the specimen and a selected-area aperture is used to limit the diffracting volume. SAED patterns are often used to determine the Bravais lattices and lattice parameters of crystalline materials by a method similar to the one used in the X-ray diffraction.

1.4.4 Ultraviolet and Visible Spectroscopy

Ultraviolet and visible (UV–Vis) absorption spectroscopy is the measurement of the attenuation of a beam of light after it passes through or reflects from a sample surface. Absorption measurements can be collected at a single wavelength or over an extended spectral range. The concentration of an analyte in solution can be determined by measuring the absorbance at a particular wavelength and applying the Beer–Lambert Law. UV–Vis range spans the range of human visual acuity between 400 and 750 nm, and deals with the study of electronic transitions between orbital's or bands of atoms, ions, or molecules in gaseous, liquid, and solid state. Therefore UV–Vis spectroscopy can be used to characterize the absorption, transmission, and reflectivity of a variety of materials, such as pigments, proteins, DNA, coatings, windows, filters, and metallic nanoparticles. Metallic and semiconductor nanoparticles depending on their size and shape are known to exhibit characteristic colors due to surface plasmon resonance. Mie, for the first time, explained the origin of color by solving Maxwell's equation for the absorption and scattering of electromagnetic radiation by small metallic particles [2]. The absorption of electromagnetic radiation by metallic nanoparticles originates from the coherent oscillation of the valence band electrons induced by an interaction with the electromagnetic field. These resonances are known as surface plasmons and occur only with nanoparticles. Hence, UV–Vis spectroscopy can be an ideal analytical tool to study the unique optical properties of nanoparticles.

1.4.5 Dynamic Light Scattering and Zeta Potential Measurements

Dynamic light scattering (DLS), also known as Photon Correlation Spectroscopy (PCS) or Quasi-Elastic Light Scattering (QELS), is a non-invasive, well-established technique commonly used to measure the size of molecules and particles typically in the submicron region, and with the latest technology lower than a nanometer. Shining a monochromatic light beam, such as a laser, onto a suspension with spherical particles in Brownian motion causes a Doppler Shift when light hits the moving particles, changing the wavelength of the incoming light. Analysis of these intensity fluctuations yields the velocity of the Brownian motion and hence the particle size using the Stokes–Einstein relationship. It is possible to compute the sphere size distribution and give a description of the particle's motion in the medium, measuring the diffusion coefficient of the particle using the autocorrelation function. The translational diffusion coefficient will depend not only on the size of the particle "core" but also on the surface structure, concentration as well as type of ions in the medium. This means that the size can be larger than measured by electron microscopy, for example, when the particle is removed from its native environment. Moreover, using DLS several other interesting parameters such as the molecular weight, radius of gyration, and translational diffusion constant can also be measured.

Almost all particulate or macroscopic materials in contact with a liquid acquire an electronic charge on their surface. Zeta potential is an important and useful indicator of this charge that can be used to predict and control the stability of colloidal suspensions. It uses electrophoretic light scattering and the laser Doppler velocimetry (LDV) method to determine particle velocity and, from this, the zeta potential. The zeta potential is the overall charge of the particles acquired in a specific medium. The magnitude of the zeta potential gives an indication of the potential stability of the colloidal system. If all the particles have a large negative or positive zeta potential they will repel each other and there will be dispersion stability and, if the particles have low zeta potential values then there will be no force to prevent the particles coming together and thereby dispersion instability.

1.4.6 Fluorescence Spectroscopy

Fluorescence spectroscopy is a spectrochemical technique of analysis, where an incident light of a fixed wavelength is directed onto the specimen prompting the transition of electron from the ground state to higher energy levels or the excited state. The molecule in the excited state then undergoes a non-radiative internal relaxation and the excited electron moves to a more stable excited level. After a characteristic lifetime in the excited state, the electron returns to the ground state by emitting photons, and in the process emit a characteristic wavelength in the form of light. This emitted energy can be used to obtain qualitative and quantitative information such as chemical composition, structure, impurities, kinetic process, and energy transfer of a material.

1.4.7 Fourier Transform Infrared Spectroscopy

Fourier transform infrared (FTIR) spectroscopy is mostly useful for identifying molecules that are either organic or inorganic. It can also be utilized to quantitate some components of an unknown mixture and for the analysis of various forms of solids, liquids, and gasses. FTIR involves the vibration of chemical bonds of a molecule at different frequencies depending on the elements and types of bonds. After absorbing electromagnetic radiation, the bond frequency increases leading to transition between ground and excited states. These absorption frequencies represent excitations of vibrations of the chemical bonds and are specific to the type of bond and the group of atoms involved in the vibration. The energy of these frequencies is in the infrared region $(4,000-400 \text{ cm}^{-1})$ of the electromagnetic spectrum. The term Fourier transform refers to a recent development wherein the data obtained in the form of an interference pattern is converted to an infrared absorption spectrum, similar to that of a molecular "fingerprint".

1.4.8 X-ray Photoelectron Spectroscopy

X-ray photoelectron spectroscopy (XPS) also known as electron spectroscopy for chemical analysis (ESCA) is the most widely used surface analytical technique because of its relative simplicity in use and data interpretation. XPS is widely used for probing the electronic structures of atoms, molecules and condensed matter. When an X-ray photon of energy (hv) is incident on a solid matter, the kinetic energy (E_k) and the binding energy (E_b) of the ejected photoelectrons can be related as $E_k = hv - E_b$. This kinetic energy distribution of the photoelectrons forms a series of discrete bands, which symbolizes the electronic structure of the sample. The core level binding energies of all the elements (except H and He) in different oxidation states are unique which provides information about the chemical composition of the sample. However, to account for the multiplet splitting and satellites accompanying the photoemission peaks, the photoelectron spectra should be interpreted in terms of many-electron states of the final ionized state of the sample, rather than the occupied one-electron state of the neutral species.

1.4.9 Atomic Absorption Spectrometry

Atomic absorption spectroscopy (AAS) uses the absorption of light to measure the concentration of gas-phase atoms. Since samples are usually in liquid or solid forms, the analyte atoms or ions must be vaporized in a flame or graphite furnace. The atoms absorb ultraviolet or visible light and make transitions to higher electronic energy levels. The analyte concentration is determined from the amount of absorption. The principle of atomic absorption spectroscopy is based on energy absorbed during transitions between electronic energy levels of an atom. When energy is provided to an atom in ground state at a high temperature (2,100–2,800°C), outer-shell electrons are promoted to a higher energy excited state. The absorbed radiation as a result of this transition between electronic levels can be used for the quantitative analysis of metals and metalloids present on solid matrices in suspension. The basis for the quantitative analysis depends on the measurement of radiation intensity and the assumption that radiation absorbed is proportional to atomic concentration. Analogy of relative intensity values for reference standards is also used to determine elemental concentrations.

1.4.10 Atomic Force Microscopy

Atomic force microscope (AFM) is a form of scanning probe microscopy (SPM) wherein a small probe is scanned across the sample to obtain information about the sample's surface. The information gathered by such interaction can be as simple as physical topography or as diverse as measurements of the material's physical, magnetic, or chemical properties. AFM has the advantage of imaging almost any type of surface such as thick film coatings, ceramics, composites, glasses, synthetic and biological membranes, microorganisms and cells, biomaterials, metals, and polymers. AFM is being applied to study various phenomena such as the abrasion, adhesion, cleaning, corrosion, etching, friction, lubrication, plating, and polishing. The AFM probe has a very sharp tip, often less than 100 Å diameter, at the end of a small cantilever beam. The probe is attached to a piezoelectric scanner tube. Interatomic forces between the probe tip and the sample surface cause the cantilever to deflect as the sample's surface topography or other properties change. A laser light reflected from the back of the cantilever measures the deflection of the cantilever. This information is fed back to a computer, which generates a map of topography and/or other properties of interest. However, based on the type of application different operation modes of AFM are used like the contact mode, where the AFM measures the hard-sphere repulsion forces between the tip and the sample; non-contact mode, where the AFM derives topographic images from measurements of attractive forces; the tip does not touch the sample, and the tapping mode, where the cantilever is driven to oscillate up and down at near its resonance frequency by a small piezoelectric element mounted in the AFM tip holder similar to non-contact mode.

1.5 Applications of Nanomaterials

The novel properties of nanoparticles hold enormous potential for their applications in several basic as well as applied sciences. Nanoscience and nanotechnology together with biology provides an excellent tool for the development of novel advanced functional systems, which can be implemented in various medical, diagnostic, and electronic applications. Advances in the field of nanotechnology

have resulted in the wide range of applications such as in electronics, communi-
cation devices, automobiles, cosmetics, paints, sensors, biological systems, and
several other consumer products. Biomedical applications include the fluorescent
biological labeling, drug and gene delivery, detection of pathogens, destruction of
pathogens, single molecule detection, protein tracking, probing DNA structure,
tissue engineering, photodynamic therapy, bio-sensors, purification and separation
of bio-molecules and cells, etc. [2, 24, 25]. A schematic illustration on the various
out reaching applications of engineered nanoparticles is described in Fig. 1.2.
These approaches represent just a few examples; it is hard to focus on all the
applications in this brief chapter; therefore, this Chapter will emphasize on the
biotechnology and biomedical applications. The applications with respect to
the use of nanoparticles for imaging and drug delivery are discussed in the
following sections, whereas the use of nanoparticles as antimicrobial agents is
discussed later in Chap. 3.

1.5.1 Nanoparticles in Imaging

There has been a remarkable progress in the utilization of novel imaging tools to
identify and monitor cells in vitro and in vivo [26]. The ability to visualize and
monitor cells morphology and function in real time for a living organism can
reveal so much information about a cells biomachinery. Cell and organism
imaging is an important criteria to understand various aspects in biology or
medicine. Different types imaging molecules are available with better sensitivity
and resolution. However, most widely used and affordable imaging technique is
the optical imaging that includes UV–Vis, fluorescence, and bioluminescence
detection. This type of detection method has been improved lately due to the
development of new fluorescent detection probes called the "quantum dots". The
elemental components in quantum dots are from the group II–VI, III–V, or IV–IV
in the periodic table, with size distributions usually less than 10 nm. Quantum dots
(QDs) are semiconductor nanoparticles with unique optical properties, properties
that lack in conventional dyes and fluorescent proteins [26–28]. Depending on the
size, quantum dots can emit light in different wavelengths than the next larger size
upon excitation. Their high stable nature and multicolor fluorescence emission
ability make them remarkable vehicles for both in vivo and in vitro imaging at
both cellular and molecular levels. Quantum dots modified using robust fluorescent
probes can also be applied to detect various cellular targets [27].

Functionalized quantum dots in conjunction with antibodies and immuno-
globulin's can be used to detect antigens on fixed and/or live cells based on
fluorescence analytical techniques. As an example, cellular uptake of quantum dots
by the plant pathogenic fungus, Aspergillus, is shown in Fig. 1.3. The advantages
of using quantum dots in imaging is their longer stability or lower photo bleaching,
which enables them to be used for longer durations, and such property lacks in
normal fluorescent dyes. Other advantages of quantum dots include their optical

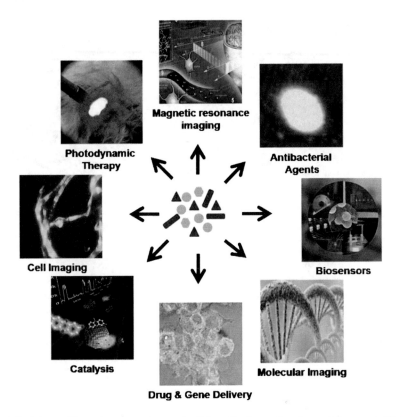

Fig. 1.2 Scheme illustrating the various potential applications of engineered nanocrystallites

properties, brighter signal intensity, narrow emission spectrum, and distinctive labeling of multiple entities. Another interesting application of quantum dots is their use in optical coding of live cells for drug screening, multiplexed assays, and their ability to track specific cell in a population of cells [29].

Though semiconductor QDs are the most widely used nanoparticles in biolabeling, issues related to their safety concerns are the major drawbacks because it has also been reported that quantum dots do elicit toxic effects to both prokaryotic and eukaryotic cell systems, either through reactive oxygen species generation or release of lethal highly toxic ions (Cd, Se, Te) [30]. Therefore, attempts were also made to exploit the use of surface plasmon resonance (SPR) properties of metallic nanoparticles for various optical imaging and labeling applications [31]. Most of the imaging tools desire expensive and sensitive lasers, components, and detectors. On the other hand use of metal nanoparticles requires a simple optical microscope equipped with a dark field condenser. Stand out example is the use gold nanoparticles, when excited by a broad white light their SPR is strongly scattered making them amenable to various biolabeling and drug delivery experiments.

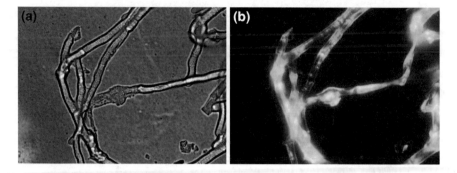

Fig. 1.3 Bright field (**a**) and the respective fluorescence image (**b**) of plant pathogenic fungus *Aspergillus* treated with cadmium sulfide quantum dots. Cell uptake of quantum dots can be clearly seen

The utilization of gold nanoparticles in biology started with the report by Mirkin's group for the optical detection of DNA [32]. Other examples include the recognition and detection of specific DNA sequences and single-base mutations in a homogeneous format. Recent application is the successful fabrication of label-free biochip, for the detection of biomolecules such as streptavidin–biotin, and was proven to be size dependent, wherein ~ 39 nm gold nanoparticles showed 20-fold better sensitivity than that of ~ 13 nm sized particles [33]. Nevertheless, it is reasonable to infer that size and shape distributions, along with the other interesting properties of gold nanoparticles might be useful to reach the different aims, as they improve biological detection sensitivity. El-Sayed et al. diagnosed cancer by imaging the cancer-specific biomarker epidermal growth factor receptor (EGFR) that was significantly produced in higher amounts on cancer cells [34, 35]. Gold nanoparticles functionalized to anti-EGFR antibodies were selectively targeted onto cancer cells to differentiate them from normal cells, which were then visualized using the dark field microscope based on gold nanoparticles surface plasmon resonance [36]. However, this diagnostic strategy is considered general, as gold nanoparticles can be easily functionalized to a range of molecules, drugs, and antibodies depending on the target agent [36]. Similarly, iron oxide nanoparticles are also being considered for imaging applications, due to several reasons; they support diagnostic imaging using CT, MRI, and PET; are highly biodegradable; are not toxic; are amenable for functionalizations; have prolonged vascular circulation time and most importantly are FDA approved [37, 38]. Iron oxide nanoparticles have been used as probes for imaging several cellular and subcellular activities. The first cellular imaging studies were performed with non-functionalized iron oxide nanoparticles for labeling leukocytes, lymphocytes, etc. [38, 39]. Various antibodies or fragments with specificity to several types of cell receptors such as HER2/Neu, LHRH, EGFR, myosine, lymphocyte, selectin, and V-CAM1 have been conjugated with iron oxide nanoparticles for both in vitro or in vivo recognition, imaging, and assessments. Examples being, the use of fluorescent magnetic nanoparticles as sensitive indicators for imaging atherosclerosis and apoptosis in vivo, upon functionalizations with vascular adhesion molecule-1 (VCAM-1) [38, 39] and annexin V [39], respectively. Similarly, targeting of the endothelial inflammatory adhesion molecule E-selectin was shown using superparamagnetic nanoparticles using both in vitro and in vivo inflammation model systems [39]. This is just a brief summary on some recent applications of nanoparticles for imaging; however, for further details the readers can refer several previously published review articles and perspectives [25, 26, 40, 41].

1.5.2 Nanoparticles in Drug Delivery

The development of engineered nanoparticles with specificity and selectivity toward the target for self-sustained drug delivery is currently an area of intense research with immense potential to revolutionize various disease treatments including cancer. With the aid of nanotechnology it might be possible to improve loopholes in therapeutics, such as the delivery of poorly hydrophilic drugs, targeted delivery of drugs selectively and specifically to cell or tumor sites, co-administer multiple drugs, and the visualization and monitoring of drug delivery site and drug action by combining novel nanotherapeutic agents together with advanced imaging techniques. Several factors can contribute to the usefulness of engineered nanoparticles in drug delivery and drug targeting systems. Major factors include the (1) Selective targeting; nanoparticles retain the capability to deliver the drugs to the vicinity of tumor due to enhanced retention time and permeability. (2) Nanoparticles tend to reduce the side effects caused by various drugs, by reducing the drug exposure on healthy tissues. This is the most promising area where nanotechnology has been implemented with potential improvements in diagnostics. The therapeutics of many currently used drugs can be improved if they can be delivered efficiently to the target; with proper use of nanotechnology. The offset of clinical implementation of current drugs is their solubility, most of the commonly used drugs are hydrophobic that makes them nonamenable for administration. The only solution for this is to make a drug hydrophilic, which can be done by functionalizing them to nanocrystallites and in turn stabilizing those using surfactants, and functionalizing them to nanoparticles that keep them in circulation for prolonged durations. Another drawback is difficulty in crossing the blood–brain barrier, especially in case of tumors, which can be overcome by drug-loaded nanoparticles, which not only enables the drug to cross the barrier but also increases its therapeutic concentrations.

Another good strategy to enhance the efficacy and decrease the toxicity of a drug is by directing the drug toward the target and retains its concentration at the specific site for sufficient time for the drug to be effective. Liposomes are excellent carriers for a majority of drugs that are used in both drug and gene delivery. Lipid cationic nanoparticles, coupled to an integrin-targeting ligand, were shown to deliver genes selectively to angiogenic blood vessels of tumor. The directed

Fig. 1.4 Confocal
microscopy image showing
the delivery of anticancer
drug doxorubicin in HEK293
cells by gold nanoparticles.
Green shows the Au-
doxorubicin conjugates
fluorescence upon uptake

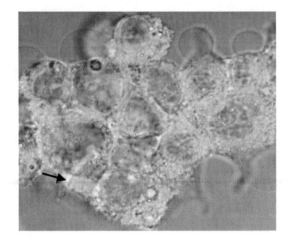

nanoparticle causes apoptosis in the tumors and sustain regression of primary and metastatic tumors.

It is also shown that drug-encapsulated nanoparticles have several fold increase in binding as compared to that of a drug alone. When it comes to drug release, gold nanoparticles also seem to be equally important, because of their wide versatility, inert, and non-toxic nature. Moreover, their photophysical properties could trigger drug release at remote place and render drug reservoirs for controlled and sustained release of drug to maintain the drug level with the therapeutic window. Gold nanoparticles could be, in principle, conjugated with several drug molecules. For example, ~ 70 molecules of paclitaxel, a chemotherapeutic drug was coupled to a ~ 2 nm gold nanoparticle [33]. Similar biogenic ~ 2–5 nm gold nanoparticles were functionalized to ~ 70–80 molecules of doxorubicin, an anticancer drug for effective HEK293 cell uptake by and toxicity [33] (see Fig. 1.4). Also the development of novel strategies for the controlled release of drugs will provide nanoparticles with the capability to deliver two or more therapeutic agents. Gold nanoparticles upon surface functionalization with polyethylene glycol (PEG) are known to have resulted in longer circulation. For example, PEGylated gold nanoparticles conjugated with tumor necrosis factor (TNF-α), a cytokine with potential anticancer effect, was proven to have increased tumor damage and reduced systemic toxicity [42]. Similarly, methotrexate, an inhibitor of dihydro-folate reductase and a chemotherapeutic agent upon functionalization with gold nanoparticles was shown to have anti-tumor effects both in vivo and in vitro against lung carcinoma [43]. Sokolov et al. showed the molecular targeting of cancer cells and tissues by functionalizing gold nanoparticles with anti-epidermal growth factor receptor (anti-EGFR) antibodies [36]. Gold nanoparticles upon conjugation to anti-HER2 antibody were proved to be a potential candidate for combinatorial therapy with imaging and hyperthermia [44]. Information on the use gold nanoparticles in various drug delivery applications is very well described in the literature [45].

Along with metal nanoparticles, metal oxide nanocrystallites are also being considered as efficient drug delivery vehicles [46]. For example, the use of ZnO quantum dots functionalized with anticancer drug, doxorubicin, has proven to be an efficient drug delivery platform for rapid and controlled release of the drug in vitro [46]. Similarly, iron oxide nanoparticles have been proven successful for delivering increased amounts of hydrophobic anticancer drug to mediate concentration-dependent anti-proliferative effects in breast and prostate cancer cell lines [46]. Iron oxide nanoparticles have also been used for the targeted delivery of therapeutic agents by functionalization to a chemotherapeutic agent such as methotrexate and chlorotoxin. These multifunctional nanoparticles showed enhanced cytotoxicity to tumor cells and prolonged tumor retention in vivo. Similarly, cerium oxide nanoparticles loaded with carboxybenzenesulfonamide have also been shown to inhibit human enzyme, carbonic anhydrase, a metalloenzyme known to be associated with glaucoma and a major cause of blindness [46]. Therefore, the relative bioavailability and biocompatibility nature of engineered metal and metal oxide nanomaterials along with their ease ability to functionalize them with various targeting molecules make them potential candidates for several drug delivery systems.

1.6 Summary

Size- and shape-dependent unique properties of nanoparticles are highly promising and bond nanotechnology with rest of the science disciplines for a better technology. Progress in nanobiotechnology or nanomedicine with the advent of novel functional materials allows us to incorporate biomolecules with nanostructured materials. Biggest advantages are likely to involve in biomedical engineering, medicine, and smart devices. Interdisciplinary approaches with the merge of different fields of science might provide a powerful synergy that takes us to a higher level of understanding and implementation of science and nanotechnology eventually leading to new innovations ultimately, advancements for the benefit of mankind.

References

1. Mirkin CA (2005) The beginning of a small revolution. Small 1(1):14–16
2. Kumar SA, Khan MI (2010) Heterofunctional nanomaterials: fabrication, properties and applications in nanobiotechnology. J Nanosci Nanotechnol 10(7):4124–4134
3. Roucoux A, Schulz J, Patin H (2002) Reduced transition metal colloids: a novel family of reusable catalysts? Chem Rev 102(10):3757–3778
4. Bhattacharya R, Mukherjee P (2008) Biological properties of "naked" metal nanoparticles. Adv Drug Deliv Rev 60(11):1289–1306

5. Kimling J, Maier M, Okenve B, Kotaidis V, Ballot H, Plech A (2006) Turkevich method for gold nanoparticle synthesis revisited. J Phys Chem B 110(32):15700–15707
6. Caswell KK, Bender CM, Murphy CJ (2003) Seedless, surfactantless wet chemical synthesis of silver nanowires. Nano Lett 3(5):667–669
7. Sun YG, Mayers B, Xia YN (2003) Transformation of silver nanospheres into nanobelts and triangular nanoplates through a thermal process. Nano Lett 3(5):675–679
8. Sun YG, Xia YN (2002) Shape-controlled synthesis of gold and silver nanoparticles. Science 298(5601):2176–2179
9. Lee KY, Hwang J, Lee YW, Kim J, Han SW (2007) One-step synthesis of gold nanoparticles using azacryptand and their applications in SERS and catalysis. J Colloid Interf Sci 316(2):476–481
10. Yang GW (2007) Laser ablation in liquids: applications in the synthesis of nanocrystals. Prog Mater Sci 52(4):648–698
11. Widiyastuti W, Balgis R, Iskandar F, Okuyama K (2010) Nanoparticle formation in spray pyrolysis under low-pressure conditions. Chem Eng Sci 65(5):1846–1854
12. Jin RC, Cao YC, Hao EC, Metraux GS, Schatz GC, Mirkin CA (2003) Controlling anisotropic nanoparticle growth through plasmon excitation. Nature 425(6957):487–490
13. Roorda S, van Dillen T, Polman A, Graf C, van Blaaderen A, Kooi BJ (2004) Aligned gold nanorods in silica made by ion irradiation of core-shell colloidal particles. Adv Mater 16(3):235–237
14. Wang ZL, Mohamed MB, Link S, El-Sayed MA (1999) Crystallographic facets and shapes of gold nanorods of different aspect ratios. Surf Sci 440(1–2):L809–L814
15. Esumi K, Matsuhisa K, Torigoe K (1995) Preparation of rod like gold particles by UV-irradiation using cationic micelles as a template. Langmuir 11(9):3285–3287
16. Kim YS, Seo JH, Cha HJ (2003) Enhancement of heterologous protein expression in *Escherichia coli* by co-expression of nonspecific DNA-binding stress protein, Dps. Enzyme Microb Technol 33(4):460–465
17. Murphy CJ, San TK, Gole AM, Orendorff CJ, Gao JX, Gou L, Hunyadi SE, Li T (2005) Anisotropic metal nanoparticles: synthesis, assembly, and optical applications. J Phys Chem B 109(29):13857–13870
18. Gole A, Murphy CJ (2004) Seed-mediated synthesis of gold nanorods: role of the size and nature of the seed. Chem Mater 16(19):3633–3640
19. Busbee BD, Obare SO, Murphy CJ (2003) An improved synthesis of high-aspect-ratio gold nanorods. Adv Mater 15(5):414–416
20. Pileni MP, Lisiecki I (1993) Nanometer metallic copper particle synhtesis in reverse micelles. Colloids Surf A Physicochem Eng Aspects 80(1):63–68
21. Rosi NL, Mirkin CA (2005) Nanostructures in biodiagnostics. Chem Rev 105(4):1547–1562
22. Zheng XW, Zhu LY, Yan AH, Wang XJ, Xie Y (2003) Controlling synthesis of silver nanowires and dendrites in mixed surfactant solutions. J Colloid Interf Sci 268(2):357–361
23. Sastry M, Swami A, Mandal S, Selvakannan PR (2005) New approaches to the synthesis of anisotropic, core-shell and hollow metal nanostructures. J Mater Chem 15(31):3161–3174
24. Suresh AK, Pelletier DA, Wang W, Broich ML, Moon J-W, Gu B, Allison DP, Joy DC, Phelps TJ, Doktycz MJ (2011) Biofabrication of discrete spherical gold nanoparticles using the metal-reducing bacterium *Shewanella oneidensis*. Acta Biomater 7:2148–2152
25. Parak WJ, Gerion D, Pellegrino T, Zanchet D, Micheel C, Williams SC, Boudreau R, Le Gros MA, Larabell CA, Alivisatos AP (2003) Biological applications of colloidal nanocrystals. Nanotechnology 14(7):R15–R27
26. Michalet X, Pinaud FF, Bentolila LA, Tsay JM, Doose S, Li JJ, Sundaresan G, Wu AM, Gambhir SS, Weiss S (2005) Quantum dots for live cells, in vivo imaging, and diagnostics. Science 307(5709):538–544
27. Medintz IL, Uyeda HT, Goldman ER, Mattoussi H (2005) Quantum dot bioconjugates for imaging, labelling and sensing. Nat Mater 4(6):435–446
28. Smith AM, Gao XH, Nie SM (2004) Quantum dot nanocrystals for in vivo molecular and cellular imaging. Photochem Photobiol 80(3):377–385

29. Delehanty JB, Boeneman K, Bradburne CE, Robertson K, Medintz IL (2009) Quantum dots: a powerful tool for understanding the intricacies of nanoparticle-mediated drug delivery. Expert Opin Drug Deliv 6(10):1091–1112

30. Hardman R (2006) A toxicologic review of quantum dots: toxicity depends on physicochemical and environmental factors. Environ Health Perspect 114(2):165–172

31. Ghosh P, Han G, De M, Kim CK, Rotello VM (2008) Gold nanoparticles in delivery applications. Adv Drug Deliv Rev 60(11):1307–1315

32. Mirkin CA, Letsinger RL, Mucic RC, Storhoff JJ (1996) A DNA-based method for rationally assembling nanoparticles into macroscopic materials. Nature 382(6592):607–609

33. Kumar SA, Peter YA, Nadeau JL (2008) Facile biosynthesis, separation and conjugation of gold nanoparticles to doxorubicin. Nanotechnology 19:495101–495111

34. El-Sayed IH, Huang X, El-Sayed MA (2006) Selective laser photo-thermal therapy of epithelial carcinoma using anti-EGFR antibody conjugated gold nanoparticles. Cancer Lett 239(1):129–135

35. Jain PK, Huang X, El-Sayed IH, El-Sayed MA (2008) Noble metals on the nanoscale: optical and photothermal properties and some applications in imaging, sensing, biology, and medicine. Acc Chem Res 41(12):1578–1586

36. Sokolov K, Follen M, Aaron J, Pavlova I, Malpica A, Lotan R, Richards-Kortum R (2003) Real-time vital optical imaging of precancer using anti-epidermal growth factor receptor antibodies conjugated to gold nanoparticles. Cancer Res 63(9):1999–2004

37. Sun C, Lee JSH, Zhang MQ (2008) Magnetic nanoparticles in MR imaging and drug delivery. Adv Drug Deliv Rev 60(11):1252–1265

38. Kievit FM, Zhang M (2011) Surface engineering of iron oxide nanoparticies for targeted cancer therapy. Acc Chem Res 44(10):853–862

39. Tassa C, Shaw SY, Weissleder R (2011) Dextran-coated iron oxide nanoparticles: a versatile platform for targeted molecular imaging, molecular diagnostics, and therapy. Acc Chem Res 44(10):842–852

40. Latorre M, Rinaldi C (2009) Applications of magnetic nanoparticles in medicine: magnetic fluid hyperthermia. P R Health Sci J 28(3):227–238

41. Sanvicens N, Marco MP (2008) Multifunctional nanoparticles - properties and prospects for their use in human medicine. Trends Biotechnol 26(8):425–433

42. Visaria RK, Griffin RJ, Williams BW, Ebbini ES, Paciotti GF, Song CW, Bischof JC (2006) Enhancement of tumor thermal therapy using gold nanoparticle-assisted tumor necrosis factor-alpha delivery. Mol Cancer Ther 5(4):1014–1020

43. Chen Y-H, Tsai C-Y, Huang P-Y, Chang M-Y, Cheng P-C, Chou C-H, Chen D-H, Wang C-R, Shiau A-L, Wu C-L (2007) Methotrexate conjugated to gold nanoparticles inhibits tumor growth in a syngeneic lung tumor model. Mol Pharm 4(5):713–722

44. Van de Broek B, Devoogdt N, D'Hollander A, Gijs HL, Jans K, Lagae L, Muyldermans S, Maes G, Borghs G (2011) Specific cell targeting with nanobody conjugated branched gold nanoparticles for photothermal therapy. Acs Nano 5(6):4319–4328

45. Sau TK, Rogach AL, Jackel F, Klar TA, Feldmann J (2010) Properties and applications of colloidal nonspherical noble metal nanoparticles. Adv Mater 22(16):1805–1825

46. Rasmussen JW, Martinez E, Louka P, Wingettt DG (2010) Zinc oxide nanoparticles for selective destruction of tumor cells and potential for drug delivery applications. Expert Opin Drug Deliv 7(9):1063–1077

Chapter 2
Facile Green Biofabrication of Nanocrystallites

Abstract To facilitate widespread applications of engineered nanoparticles, researchers are looking at novel and better synthesis strategies. This brief chapter details the biofabrication/biosynthesis of nanoparticles, advantages of biofabrication over conventional chemical or physical routes of synthesis, and the mechanism involved in biofabrication. An illustration on the biosynthesis of silver nanoparticles using the metal-reducing bacterium *S. oneidensis* will be presented as an example. Further, details on the synthesis methodology, physical characterizations with respect to morphology, crystallinity, surface properties, and size and shape distributions, which will be based on characterization techniques involving UV–Vis and Fourier transform infrared spectroscopy, dynamic light scattering, X-ray diffraction, transmission electron microscopy, and atomic force microscopy measurements will be discussed.

Keywords Biofabrication · Nanocrystallites · Microbial · Mechanism

2.1 Biofabrication

Naturally occurring biological entities such as proteins or peptides, nucleic acids, nucleotides, and plant extracts have been exploited for the production of hierarchically assembled advanced nanomaterials lately. The involvement of biological entities in the growth and nucleation of nanocystallites are gaining tremendous interests for several reasons, most importantly anisotropic metallic structures with size and shape control can be produced, are cost-effective, green, and energy intensive. Different types of engineered nanocrystallites with specific dimensions, shape distributions, forms, and hierarchies have been produced utilizing various biomimetic templates. Fabrication of one-dimensional parallel and two-dimen-

A. K. Suresh, *Metallic Nanocrystallites and their Interaction with Microbial Systems*, SpringerBriefs in Biometals, DOI: 10.1007/978-94-007-4231-4_2, © The Author(s) 2012

sional crosses of palladium nanowire arrays, copper nanowires, and silver nano-rods were performed on a solid substrate surface templated by DNA followed by reduction of metal ions [1]. It was opined that the formation of unique crystalline nanorods and nanowires arise due to template directed aggregation of small par-ticles and its subsequent re-crystallization, rather than simple agglomeration. Similarly, self-assembled two-dimensional DNA grids were used as templates to grow 5 nm gold nanoparticles into periodic square lattices that have applications in nanospectronic and nanophotonic devices [2]. Silver nanoparticles, nanorods, and nanowires were synthesized by initial complexation of silver with DNA and then reducing the complex with sodium borohydride. CdS quantum dots of narrow size distributions (6 nm) were also fabricated using wild-type tRNA and an unfolded mutant tRNA of similar lengths. One type of DNA molecule was reported to mediate the nucleation and growth of $CaCO_3$ particles of diverse morphologies [2]. The authors suggested that the concentration of DNA significantly influenced the shape distributions of the particles. These and several other demonstrations clearly reveal that biological molecules can definitely have an impact on nanocrystallites size and shape distribution. Under appropriate conditions different segments of the block copolymers have been shown to form regular arrays of cylinders with structures similar to the surfactant used [3]. Different regions of arrayed structures could be designed by selective interaction of functional groups with the precursor metal ions through physical adsorption or chemical bonding followed by its reduction resulting in the formation of one-dimensional nanostructures. By using a wide range of copolymers this templating procedure has also been exploited for the synthesis of silver nanowires, and gold nanowires and nanosheets. Capillaries of single-walled carbon nanotubes have been used for the synthesis of nanowires of gold, silver, platinum, and palladium [4]. Polystyrene mesospheres were also used as templates for the synthesis of silver nanodisks [2]. Silver nanocubes and nanotriangles themselves have also been utilized for the synthesis of nanoboxes and triangular rings of gold [5]. In this approach the synthesis is facilitated by the transmetallation of gold ions by metallic silver.

Enzymes, nature's catalysts, and peptides provide functional building blocks for the development of advanced materials and tend to perform reactions much faster, under mild conditions in a highly specific manner. The oxidation/reduction mechanism involved in the formation of nanomaterials coupled with the self-assembling ability of enzymes to carry out such reactions have prompted inves-tigators to examine the role of proteins and enzymes in the biotransformation of metals. Engineered proteins and peptides that recognize inorganic surfaces were also proven to be successful in the generation and assembly of various inorganic nanostructures. For example, in vitro synthesis of magnetite nanocrystallites of uniform size distributions of ~ 30 nm were achieved using the Mms6 protein purified from the magnetotactic bacteria [6]. Similarly, silicateins are known to promote biosilica formation in nature; silicatein filaments have been shown to produce different forms of nanoparticles like the titanium dioxide, gallium oxohydroxide, and γ-gallium oxide in vitro [7, 8]. An enzymatic biocatalyst purified from marine sponge, *Tethya aurantia* was used as a catalyst as well as a template to hydrolyze and condense molecular precursor $BaTiF_6$ at low temperatures to produce $BaTiOF_4$ nanocrystallites [9]. In another report, standard lab protein bovine serum albumin has been shown to act as both a reducing as well as a stabilizing agent for the synthesis of various metal (Au, Ag) and alloy (Au–Ag) nanoparticles [10]. Brown using the gold binding peptides, identified from a cell surface library, synthesized very big platelets of gold nanocrystallites [11]. Similarly, nanometer-scale thick gold nanocrystalline platelets with high yields were synthesized using a protein purified from the green algae, *Chlorella vulgaris* [12]. In two independent investigations, Kumar et al. showed that enzymes, nitrate reductase, and sulfite reductase purified from the plant pathogenic fungus, *Fusarium oxysporim*, can be used for the production of silver and gold nanoparticles respectively [13, 14]. In another investigation, the authors further showed that using a similar enzyme (sulfite reductase), chemically difficult to fabricate nanoparticles of metal sulfides can also be synthesized [15]. Similarly, Liu et al. using the enzyme phytochelatin synthase and the capping agent phytochelatin purified from the model bacterium, *E. coli*, showed the synthesis of small CdS quantum dots of less than 5 nm [16]. Recently, gold nanocrystallites of various anisotropic size and shape distributions were synthesized using the dodecapeptide, Midas-2, selected from a phage-displayed combinatorial peptide library [12]. The authors further claimed that changes in the single amino acid in the peptide, which are in turn controlled by the pH, were responsible for the production of diverse shapes such as nanospheres, nanoribbons, nanowires, and nanoplatelets. For a detailed understanding of the various polypeptides used to produce different types of nanoparticles, the readers can refer the cited review article [17].

Lately, several microbial systems including fungi, yeast, and bacteria have been utilized as environmentally benign nano-factories for the production and assembly of various types, size, and shape distributions, and bimetallic alloy nanoparticles. Microorganisms have long been known to develop resistance to metal ions either by sequestering metals inside the cell or by effluxing them into the extracellular medium. The ability of microorganisms to produce inorganic materials either intra- or extracellularly has been known for more than 30 years. This unique microbial behavior has been re-exploited lately, for the production of diverse engineering nanostructures of metals, metal oxides, alloys, and other complex structures. There are several reports on the microbial-based biosynthesis of different types of nanoparticles, viz. gold, silver, platinum, palladium, magnetite, cadmium sulfide, cadmium selenide, etc. using microorganisms such as *Fusarium, Enterobacteria, Shewanella, Geobacter, Pseudomonas, Cyanobacteria, Bacillus, Escherichia,* and *Aspergillus.* In these studies biosynthesis completely relies on the microorganism's reductive capabilities and intrinsic metal resistance mechanism to quell metal ion stresses [2]. The source of inspiration for biofabrication has always been nature. Many naturally existing organisms are capable of producing inorganic structures for protection, survival, tools, or weapons for self-defense, etc. For example, mollusks produce hard shells or nacres made of crystalline calcium carbonate for protection against predators. Mangetotactic bacteria are known to produce magnetite nanoparticles, well-assembled in the subcellular membrane that

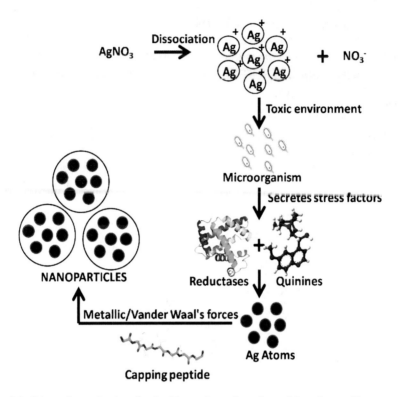

Fig. 2.1 Schematic mechanism for the biotransformation of metal ions into stable nanocrystallites by microorganisms, illustrated using the biosynthesis of silver nanoparticle by the fungus, *Fusarium oxysporum*

recognize magnetic currents for migration and alignment purposes [18]. Utilizing the biofabrication technique, and using the metal-reducing bacterium, *S. oneidensis* nearly monodispersed nanoparticles have also been produced [19].

Monodispersed nanoparticles with specific size and shape distributions are important for various applications in biological and chemical sensing, optical, medical, and electronic devices due to their unique optical features [19]. Nanomaterials created through biosynthesis have the advantages of being highly reproducible, water soluble, and environmentally benign since the use of toxic surfactants and chemicals is unnecessary. The process involves cellular secretion of NADH-dependent reductases to quell metal ion stresses. As illustrated in Fig. 2.1, demonstrating the biosynthesis of silver nanoparticles as an example, the possible synthesis mechanism might involve an enzymatic reduction of metal ions utilizing NADH-dependent reductases to convert toxic metal ions (M^+) into stable metal atoms (M^0) and subsequent stabilization using the capping protein/peptide secreted by the bacteria under metal stress. Several studies performed by independent investigators have demonstrated that the biotransformation might involve a complex of either reductases and capping peptides, quinines or cytochromes, phytochelatins or electron shuttles that are known to reduce various metal and metal oxides [20]. Apart from metallic nanoparticles, semiconductor CdS quantum dots were also synthesized utilizing the above-mentioned method using the enzyme sulfite reductase [15]. These in vitro syntheses of metallic gold, silver, and semiconductor CdS nanoparticles utilizing the enzymes confirms the mechanism proposed for the in vivo conversion of metal ions into nanoparticles by the microorganism. The defense mechanism definitely can be and has been exploited as a means for the production of diverse nanoparticles, and has overcome the disadvantages of similar chemical/physical routes of nanoparticle synthesis. Biosynthesis also avoids the use of toxic surfactants and solvents, is environmentally benign, highly reproducible, and produces hydrophilic and biocompatible nanoparticles.

Detailed below is an illustration of the process of biofabrication, monitoring, and evaluation of the biosynthesis of silver nanoparticles by the well-studied metal-reducing bacterium, *S. oneidensis* [20].

2.1.1 Biofabrication of Silver Nanoparticles

Silver nanocrystallites are among the many nanoparticles of particular interest because of their well-known bactericidal and fungicidal properties and are therefore widely used in consumer products such as medical devices, textiles, food packaging, and health care and household products. One type of silver nanomaterial from namely silver sulfadiazine is used clinically to reduce burn or wound infections caused by multi-drug resistant bacteria and fungi. Silver nanoparticles are also used in several water purification and air quality management systems. Due to the widespread use of these nanomaterials, there is ever-growing need regarding their synthesis; for large-scale production, addressing economic concerns, and most importantly environmentally benign green synthesis procedures.

2.1.1.1 Biosynthesis Methodology

The methodology illustrated here is for the production of silver nanoparticles using the metal-reducing bacterium *S. oneidensis* [20]. A single bacterial colony from an overnight grown bacterial colony in Luria–Bertani agar Petri dish was inoculated into 100 ml of the fresh growth medium, Luria–Bertani in a 500 ml Erlenmeyer flask and was incubated at 30°C on a shaker at 200 rpm for 24 h. The bacteria grown were collected by spinning down the bacteria by centrifugation (5,000×g, 20 min); the obtained bacterial cells were washed thoroughly with distilled water under sterile conditions, to get rid of the leftover nutrient medium and cellular secretions if any. In another 500 ml Erlenmeyer flask ∼3–5 g wet bacterial biomass was suspended in 100 ml aqueous solution containing 1 mm AgNO$_3$ and incubated under the same conditions mentioned above. The formation of silver nanoparticles was monitored based on visual inspection as well as by performing UV–Vis absorption measurement for 1 ml aliquots over a spectral range of 200–700 nm at regular intervals. After completion of the synthesis process (~ 48 h), the reaction mixture was first centrifuged ($5,000 \times g$, 20 min) to remove the bacteria, filtered using a sterile 0.2 μm syringe filter, to get rid of high molecular weight cellular secretions, and the particles were collected by high speed ultracentrifugation ($100,000 \times g$, 45 min). The obtained silver nanoparticles were washed a couple of times with Milli Q water and the biogenic-Ag nanoparticles were further characterized for their purity, crystallinity, surface characteristics, size and shape distributions using various advanced analytical tools and techniques as described below.

2.1.2 Characterization of the Silver Nanoparticles

The first and the foremost indication for the formation of nanoparticles is the change in its color (see Fig. 2.2a) that originates from their size- and shape-dependent surface plasmon resonance (SPR). SPR is a unique phenomenon for investigating several catalytic and oxidation/reduction reactions, sensing, alloying, and electrochemical processes. Not all nanoparticles have SPR, the very few metals with SPR include gold, silver, copper, tin, lead, mercury, cadmium, indium, and the alkali metals. The SPR varies with the nanoparticle type and therefore has its own characteristic color, for example gold; ruby red, silver; brown, cadmium; yellow, magnetite; black; platinum, brownish red. For a given particular type of nanoparticle, change in the color can also occur due to the changes in the size and/or shape distributions of the particles. Therefore, UV–Vis spectra of the same nanoparticles, with different size distributions, should reveal little shift in the surface plasmon resonance (SPR) band. UV–Visible spectra in this study were recorded on a CARY 100 Bio spectrophotometer (Varian Instruments, California) operated at a resolution of 1 nm. The SPR for small silver nanoparticles, of sizes $\sim 4 \pm 2.5$ nm, showed an intense band centered at 410 nm, which is due to the excitation of SPR in the metal nanoparticles, as can be seen in Fig. 2.2b, and suggests the presence of silver nanoparticles.

Irrespective of the methodology employed to synthesize nanoparticles, most of the nanoparticles, if not encapped by a stabilizing molecule tend to aggregate or agglomerate, due to weak Wander Wall's forces or intermetallic interactions, and if not bothered they eventually form big clumps and precipitate at the bottom, and such aggregated nanoparticles do find applications where no surface coat is required such as in thin films, electric annealing, etc. However, for most biomedical and analytical applications, highly stable and soluble nanoparticles are desired. And as described earlier, to achieve such stable nanoparticles, various researchers incorporate different additives in the form of capping agents, solvents, and templates, based on the specific application the particles are desired for.

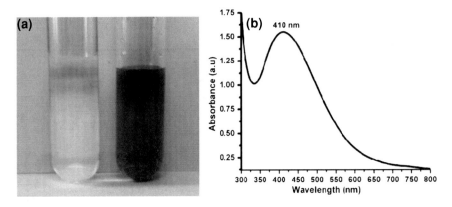

Fig. 2.2 a Image of the test tubes containing AgNO₃ solution before (test tube on the *left*) and after (test tube on the *right*) the formation of silver nanoparticles by the *S. oneidensis* biomass. **b** UV–Vis spectra of the silver nanoparticles with intense plasma peak at 410 nm

However in this case, as the particles are synthesized using microbial-based bio-synthesis, the stabilizing molecule encapping their surface is not completely understood yet. Therefore, to partially elucidate the nature of stabilizing molecule surrounding the biogenic-Ag nanoparticles produced by the microorganism, Fourier transform infrared spectroscopy measurement was performed. As mentioned earlier, FTIR is a basic analytical tool that analyses the specific functional moieties associated within the given sample. FTIR analysis of the samples dried on a ZnSe window was performed on a Nicolet Magna-IR 760 spectrophotometer at a resolution of 4 cm^{-1}. As seen in Fig. 2.3, FTIR spectroscopy for the biogenic-Ag nanoparticles revealed the presence of vibration bands centered at regions 1,057, 1,398, 1,653, 2,360, and 2,930 along with an intense broad band at 3,292 cm^{-1}. The band corresponding at 1,653 cm^{-1} arises due to the presence of carbonyl (–C–O–C– or –C–O–) stretch and –N–H stretch vibrations in the amide linkages, the small peak at 1,398 cm^{-1} was also observed from amide III, clearly indicating the involvement of protein or a peptide on the surface that likely appears to be acting as a capping/stabilizing molecule. The vibration bands at 1,057 cm^{-1} correspond to the carbonyl group and alcoholic groups respectively. The band at 3,292 cm^{-1} is characteristic of the hydroxyl functional group in alcohols and phenolic compounds. We have also demonstrated that the capping protein, which may contribute to the overall stability and integrity of the nanoparticles, can be removed from the surface of the particles by a detergent (e.g. sodium dodecyl sulfate) treatment, followed by boiling for half an hour. Such a treatment may be necessary for specific applications where a surface coat is not desired, such as in particle annealing and thin film formation.

Another important characterization technique used to assess the purity, structure, and crystallinity of the nanoparticles is the X-ray diffraction (XRD). Every type of nanocrystalline material has its own characteristic Braggs reflections or

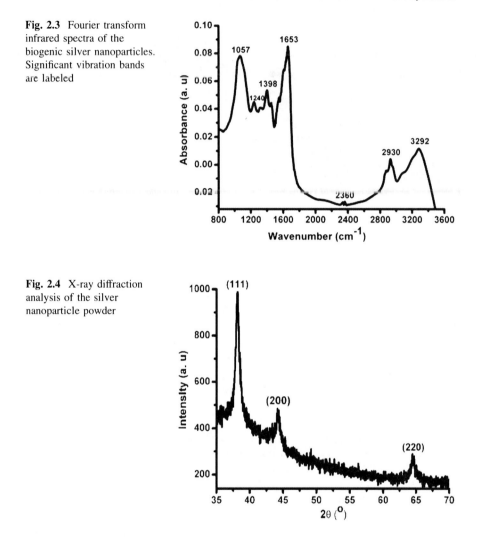

Fig. 2.3 Fourier transform infrared spectra of the biogenic silver nanoparticles. Significant vibration bands are labeled

Fig. 2.4 X-ray diffraction analysis of the silver nanoparticle powder

Bragg peaks that are documented in a book called the Methods in X-ray Crystallography often used as a standard reference. Depending on the compatibility of the XRD machine, samples for the XRD can be prepared in several ways; for example, a thick layer can be coated onto a solid Si(III) wafer; sample can be dried to powder and can be used directly. In this study, XRD of dried silver nanoparticle powder was performed on a Discover D8 X-ray diffractometer with an Xe/Ar gas-filled Hi Star area detector and an XYZ platform, operated at 40 kV and at a current of 40 mA. XRD of biogenic-Ag nanoparticle powder showed intense Bragg peaks at (111), (200), and (220) in the 2θ range of 35–70θ (see Fig. 2.4) and agree with the values that are reported for silver nanocrystals, thereby confirming the purity and crystallinity of the biogenic-Ag nanoparticles.

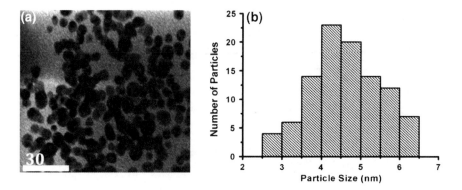

Fig. 2.5 **a** TEM image of the biogenic silver nanoparticles. **b** Histogram particle size distribution measurements made from the TEM image by counting ∼100 particles in order to obtain average particle diameter

As mentioned earlier size and shape are the most important characteristics considered for any nanomaterial before it can be implemented in an application, as the properties of nanoparticles primarily depend on their size and shape distributions, and are determined based on transmission electron microscopy measurements. Samples for the TEM measurements were prepared by drop coating the nanoparticles sample on carbon coated copper TEM grids followed by air drying the grids for a couple of hours. TEM measurements were performed on a LIBRA®120 PLUS transmission electron microscope (Carl Zeiss, Germany). TEM images of the biogenic-Ag nanoparticles revealed a large number of nearly monodispersed particles as shown in Fig. 2.5. Closer inspection of the particle morphology, at higher magnifications and different locations of the grid showed spherical single well dispersed nanoparticles as well as aggregates. An estimate of the size of the particles was also made from the line broadening of the (112) reflection pattern using Debye–Scherrer's formula ($D = 0.94$ $\lambda/b\cos\theta$), and are in good agreement with the nanoparticles size estimated by TEM analysis. The particle size distribution was also performed using dynamic light scattering measurements. However, based on dynamic light scattering the hydrodynamic sizes of the nanoparticles appeared to be much larger (∼82.5 nm) when compared to the TEM measurements. This may be attributed to overlapping particles and the electrical double layer phenomenon that occurs with charged particles which can affect DLS measurements, while TEM imaging allows latitude for eliminating aggregated particles from the analysis. A particle size histogram plot (plotted using Image J software) from the TEM image showed the size distribution of biogenic silver nanoparticles to be in the range between ∼2 and 11 nm with the largest number of particles being 4 ± 1.5 nm (see Fig. 2.5b). The particles were negatively charged with a zeta potential of -12 ± 2 mV, which can be one of the reasons for their long-term stability; the electrostatic repulsive forces between the nanoparticles might protect them from association and thereby prevent agglomeration or clumping in aqueous suspension. The dynamic light scattering and the

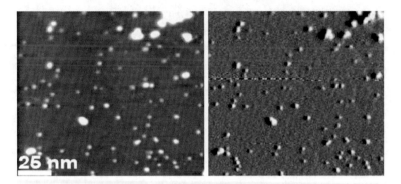

Fig. 2.6 Atomic force microscopy analysis of the biogenic silver nanoparticles. AFM topographical (*left*) and deflective (*right*) images. The deflection is shown to differentiate single particles and clumps of particles

zeta potential measurements were performed on a Brookhaven 90 Plus/BI-MAS Instrument (Brookhaven Instruments, NY).

The size and shape distribution of the nanoparticles was also confirmed by performing the atomic force microscopy (AFM) analysis. As can be seen in Fig. 2.6, AFM revealed uniformly shaped well dispersed nanoparticles with a particle height ranging from ~ 2 to 11 nm, similar to the size distribution observed in TEM measurements. The samples for the AFM analysis were prepared by coating a drop of the nanoparticle sample onto plain mica and drying overnight; the particles dispersed well on mica surface, characteristic of a hydrophilic protein surface coat. For AFM, samples were imaged in either contact or intermittent contact mode with a PicoPlus AFM (Aligent Technologies, Tempe, AZ) using a 100 μm scanning head at 128–512 pixels per line scan and a scan speed of 0.5 line/s. The cantilevers used were Veeco silicon nitride probes (MLCT-AUHW, Veeco, Santa Barbara, CA).

2.2 Summary

For the past few decades, there has been a tremendous progress in the biofabrication of engineered nanostructures of controlled morphology and size distributions at the nanometer scale. Microorganisms are best known machines that function at the nanoscale, performing numerous operations ranging from energy production to target material extraction with very high efficiency. Microorganisms have been successfully implemented for the synthesis, nucleation, and assembly of various metal, metal oxide, and metal sulfide nanocrystallities under ambient conditions. The produced particles have interesting properties like hydrophilic nature, high stability, biocompatibility, high surface area. Additionally, biological entities have also been shown to have an influence on the size and shape distributions, and surface properties of diverse nanoparticles for various intended applications. With the continued interest and ingenuity of researchers from various disciplines, the future materials promise to be green, exciting, with selectively designed nanoparticles for diverse applications.

References

1. Wei G, Wang L, Liu ZG, Song YH, Sun LL, Yang T, Li ZA (2005) DNA-network-templated self-assembly of silver nanoparticles and their application in surface-enhanced Raman scattering. J Phys Chem B 109(50):23941–23947
2. Kumar SA, Khan MI (2010) Heterofunctional nanomaterials: fabrication, properties and applications in nanobiotechnology. J Nanosci Nanotechnol 10(7):4124–4134
3. Kim JH, Lee TR (2004) Thermo- and pH-responsive hydrogel-coated gold nanoparticles. Chem Mater 16(19):3647–3651
4. Govindaraj A, Satishkumar BC, Nath M, Rao CNR (2000) Metal nanowires and intercalated metal layers in single-walled carbon nanotube bundles. Chem Mater 12(1):202–205
5. Sun SH, Anders S, Thomson T, Baglin JEE, Toney MF, Hamann HF, Murray CB, Terris BD (2003) Controlled synthesis and assembly of FePt nanoparticles. J Phys Chem B 107(23):5419–5425
6. Galloway JM, Arakaki A, Masuda F, Tanaka T, Matsunaga T, Staniland SS (2011) Magnetic bacterial protein Mms6 controls morphology, crystallinity and magnetism of cobalt-doped magnetite nanoparticles in vitro. J Mater Chem 21(39):15244–15254
7. Kisailus D, Truong Q, Amemiya Y, Weaver JC, Morse DE (2006) Self-assembled bifunctional surface mimics an enzymatic and templating protein for the synthesis of a metal oxide semiconductor. Proc Natl Acad Sci U S A 103(15):5652–5657
8. Kisailus D, Choi JH, Weaver JC, Yang WJ, Morse DE (2005) Enzymatic synthesis and nanostructural control of gallium oxide at low temperature. Adv Mater 17(3):314–318
9. Brutchey RL, Yoo ES, Morse DE (2006) Biocatalytic synthesis of a nanostructured and crystalline bimetallic perovskite-like barium oxofluorotitanate at low temperature. J Am Chem Soc 128(31):10288–10294
10. Singh AV, Bandgar BM, Kasture M, Prasad BLV, Sastry M (2005) Synthesis of gold, silver and their alloy nanoparticles using bovine serum albumin as foaming and stabilizing agent. J Mater Chem 15(48):5115–5121
11. Brown S (1997) Metal-recognition by repeating polypeptides. Nat Biotechnol 15:269–272
12. Kim J, Rheem Y, Yoo B, Chong Y, Bozhilov KN, Kim D, Sadowsky MJ, Hur H-G, Myung NV (2010) Peptide-mediated shape- and size-tunable synthesis of gold nanostructures. Acta Biomater 6(7):2681–2689
13. Kumar SA, Abyaneh MK, Gosavi SW, Kulkarni SK, Ahmad A, Khan MI (2007) Sulfite reductase-mediated synthesis of gold nanoparticles capped with phytochelatin. Biotechnol Appl Biochem 47:191–195
14. Kumar SA, Abyaneh MK, Gosavi SW, Kulkarni SK, Pasricha R, Ahmad A, Khan MI (2007) Nitrate reductase-mediated synthesis of silver nanoparticles from AgNO$_3$. Biotechnol Lett 29(3):439–445
15. Ansary AA, Kumar SA, Krishnasastry MV, Abyaneh MK, Kulkarni SK, Ahmad A, Khan MI (2007) CdS quantum dots: enzyme mediated in vitro synthesis, characterization and conjugation with plant lectins. J Biomed Nanotechnol 3(4):406–413
16. Liu F, Kang SH, Lee Y-I, Choa Y-h, Mulchandani A, Myung NV, Chen W (2010) Enzyme mediated synthesis of phytochelatin-capped CdS nanocrystals. Appl Phys Lett 97(12):123703
17. Slocik JM, Naik RR (2007) Biological assembly of hybrid inorganic nanomaterials. Curr Nanosci 3(2):117–120

18. Mann S, Frankel RB, Blakemore RP (1984) Structure, morphology and crystal-growth of bacterial magnetite. Nature 310(5976):405–407
19. Suresh AK, Pelletier DA, Wang W, Broich ML, Moon J-W, Gu B, Allison DP, Joy DC, Phelps TJ, Doktycz MJ (2011) Biofabrication of discrete spherical gold nanoparticles using the metal-reducing bacterium *Shewanella oneidensis*. Acta Biomater 7:2148–2152
20. Suresh AK, Pelletier DA, Wang W, Moon J-W, Gu B, Mortensen NP, Allison DP, Joy DC, Phelps TJ, Doktycz MJ (2010) Silver nanocrystallites: biofabrication using *Shewanella oneidensis*, and an evaluation of their comparative toxicity on Gram-negative and Gram-positive bacteria. Environ Sci Technol 44(13):5210–5215

Chapter 3
Engineered Metal Nanoparticles and Bactericidal Properties

Abstract In this chapter, the bactericidal properties of engineered metal nanoparticles will be discussed. Toxicity assessments for nanoparticles, reasons for nanoparticles being considered toxic, and the necessity to understand the toxicity for nanoparticles, various causes, and the proposed mechanisms behind nanotoxicity will be described. Also, details on the various techniques used to assess bactericidal activity, their advantages and limitations, influence of size, shape, surface coatings, along with the mechanistic of bacteria–nanoparticle interactions will be discussed.

Keywords Bactericidal · Comparative · Drug resistance · Metal nanoparticles · Toxicity

3.1 Nanoparticles as Bactericidal Agents

Outbreak of infectious diseases caused by various dreadful pathogenic microorganisms and the emergence of antibiotic-resistant pathogens such as multi-drug resistant bacteria, fungi, and parasites, led researchers and pharmaceutical companies to hunt for new and better antimicrobial agents. Therefore, the development of novel antimicrobial compounds or the advance formulations of those available for improved and sustainable antimicrobial activity for diagnosis, antisepsis, or disinfection is a high priority area of research.

Nanoparticles have been emerging as new smart materials with antimicrobial properties due to their large surface area and unique catalytic properties. Several types of nanoparticles such as copper, zinc oxide, cadmium, titanium dioxide have been proven to have bactericidal properties. Silver nanoparticles are a prime example of such material with exceptional antibacterial activity. Silver nanoparticles are proven to have antimicrobial efficacy against diverse bacteria, viruses, and other

A. K. Suresh, *Metallic Nanocrystallites and their Interaction with Microbial Systems*,
SpringerBriefs in Biometals, DOI: 10.1007/978-94-007-4231-4_3,
© The Author(s) 2012

eukaryotic microorganisms. Silver in its colloidal form is being used for the treatment of burns and chronic wounds for centuries. Some forms of silver have also been demonstrated to be effective against burn infections, severe chronic osteomyelitis, urinary tract infections, and central venous catheter infections. As early as 1000 BC silver was used to make water potable (purifying water for drinking purposes). In the 1700s, silver nitrate was used for the treatment of venereal diseases, fistulae from salivary glands, and bone and perennial abscesses. In the nineteenth century granulation tissues were removed using silver nitrate to allow epithelialization and promote crust formation on the surface of wounds. Varying concentrations of silver nitrate was used to treat fresh burns. In 1881, ophthalmia neonatorum was cured using silver nitrate eye drops. Silver impregnated dressings were designed for skin grafting. However in the 1940s, after penicillin was introduced, the use of silver for the treatment of bacterial infections was minimized. Silver again came into the picture in the 1960s when Moyer introduced the use of 0.5% silver nitrate for the treatment of burns. He proposed that this solution does not interfere with epidermal proliferation and possesses antibacterial property against diverse microorganisms such as *Staphylococcus aureus, Pseudomonas aeruginosa,* and *Escherichia coli.* In 1968s, silver nitrate was combined with sulfonamide to form silver sulfadiazine cream, which served as a broad-spectrum antibacterial agent and was used for the treatment of burns [1]. Silver sulfadiazine is effective against bacteria like *E. coli, S. aureus, Klebsiella* sp., *Pseudomonas* sp [1]. It also possessed some antifungal and antiviral activities. Recently, due to the emergence of antibiotic-resistant bacteria and limitations of the use of antibiotics clinicians have returned to silver wound dressings containing varying levels of silver. The literature suggests that silver nanoparticles have been drastically exploited in medicine for the treatment of burns and wounds, in dental materials, coating stainless steel materials, textiles, water treatment, sunscreen lotions etc. and possess lower cytotoxicity, high thermal stability, and low volatility [1].

Detailed below is an illustration of the comparative toxicity assessments of our published results [1] which demonstrate that the physical or chemical properties of the silver nanoparticles including surface charge, differential binding potential, which are in turn influenced by the surface coatings and the synthesis methodology employed to fabricate the particles, are the major determinant factors in eliciting potential bactericidal activity and in dictating silver nanoparticles' interaction with the specific bacteria. To have a better understanding on the individual contribution of these factors in eliciting bacterial toxicity, studies using three different surface engineered silver nanocrystallites with nearly uniform size and shape distribution but with different surface coatings, imparting overall high negativity to high positivity were performed. These nanoparticles included biogenic-Ag (described earlier in Chap. 2), colloidal-Ag, and oleate-Ag nanoparticles with zeta potentials -12 ± 2 mV, -42 ± 5 mV and -45 ± 5 mV respectively. The particles were purified and thoroughly characterized so as to avoid false bactericidal interpretations. Regardless of the synthesis methodology employed, all the different types of silver nanoparticles examined were less than 20 nm and had a narrow size distribution. As can be inferred from Fig. 3.1, which shows the TEM images of the

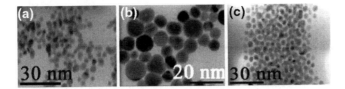

Fig. 3.1 Transmission electron microscopy images of various silver nanoparticles **a** biogenic-Ag, **b** uncoated-Ag and **c** oleate-Ag nanoparticles used for the comparative bactericidal assessments

different silver nanoparticles used for toxicity studies, colloidal-Ag nanoparticles were ~3–18 nm, an average of 8 ± 2 nm (see Fig. 3.1b), biogenic-Ag nanoparticles were ~2–11 nm, an average of 4 ± 1.5 nm (see Fig. 3.1a), and oleate-Ag nanoparticles were ~3–8 nm, an average of 4 ± 1 nm (see Fig. 3.1c). We have chosen small silver nanoparticles because of their high penetrating power and their electronic effects, and electronic effects are known to enhance the reactivity of the nanoparticles.

3.2 Bacteriological Toxicity Assessment

Analytical techniques that can assess the bactericidal activity of a material are required to improve drug formulations for the treatment of infections caused by various multi-drug resistance microorganisms. The bactericidal activity of a material can be evaluated using several toxicity assessing analytical methodologies, and are described below:

3.2.1 Disk Diffusion Tests

When a filter paper disk impregnated with a material possessing toxic properties is placed on LB agar, the material based on its diffusability tends to diffuse from the disk into the agar. This diffusion will place the material in the agar surrounding the disk, which further depends on factors such as solubility, diffusion rate, and molecular size that ultimately determine the distance of the area of materials infiltration around the disk. If a bacterium is placed on the agar it will not grow in the region around the disk if it is susceptible to the material. This area of no bacterial growth surrounding the disk is known as the "diameter of the zone of inhibition (DIZ)" and the method is called the disk diffusion test. The bacterial

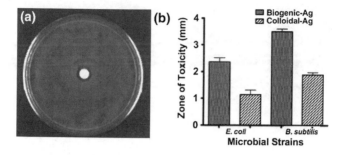

Fig. 3.2 a Image of an agar plate containing one of the silver nanoparticle-impregnated disks showing the diameter of the inhibition zone. **b** Quantitative data analysis on the zone of inhibition showed by biogenic-Ag and uncoated-Ag nanoparticles on *E. coli* and *B. subtilis*

sensitivity to different engineered silver nanoparticles was tested using disk diffusion test, for which stocks of equal concentrations (25 μg/mL) of all the different types of silver nanoparticles were first made. Then, small disks of uniform size (6 mm diameter) were placed separately in the biogenic-Ag or colloidal-Ag or oleate-Ag nanoparticle stock solutions for 5 min and the disks were removed carefully using sterile forceps. The bactericidal suspension (100 μL of 10^4–10^5 CFU mL^{-1}) was spread plated uniformly on the LB agar Petri dishes using a sterile spreader under sterile conditions, before placing the disks on the plate. The plates were then incubated at 37°C for 18 h, after which the average diameter of the inhibition zone surrounding the disks was measured using a ruler of 1 mm resolution.

When the antibacterial activity for variously synthesized silver nanoparticles on the Gram-negative (*E. coli*) and Gram-positive (*B. subtilis*) bacteria were compared using the diameter of zone of inhibition (DIZ) in disk diffusion assay, the strains susceptible to disinfectants showed larger DIZ (see Fig. 3.2a), whereas resistant strains exhibited smaller or no DIZ. The disks with biogenic-Ag were surrounded by a larger DIZ as compared to DIZ with colloidal-Ag for both Gram-negative (*E. coli*) and Gram-positive (*B. subtilis*) bacterial strains, as shown in the form of graphic illustration (see Fig. 3.2b). The DIZ of biogenic-Ag on *B. subtilis* was almost 20–30% greater than that observed for the *E. coli*. Similarly, for *B. subtilis* the colloidal-Ag was found to be more effective when compared to *E. coli*, however the difference in the DIZ was merely 10–15%, whereas, oleate-Ag showed almost no DIZ for both *B. subtilis* and *E. coli* strains that were analyzed. This method illustrates the potential enhancement in the bactericidal activity of biogenic-Ag as well as colloidal-Ag on both *E. coli* and *B. subtilis* bacterial strains. As the DIZ measurements were performed on agar Petri dishes using a ruler with a resolution of 1 mm, the possibility of error exists; however, the method illustrates the potential enhancement in the bactericidal activity of biogenic silver nanoparticles to both Gram-negative (*E. coli*) and Gram-positive (*B. subtilis*) bacterial strains.

Table 3.1 Minimum inhibitory concentrations of the various types of silver nanoparticles on *E. coli*, *B. subtilis*, and *S. oneidensis*

Nanoparticle type	Minimum inhibitory concentration (µg/mL)		
	E. coli	*B. subtilis*	*S. oneidensis*
Biogenic-Ag	2	0.5	3
Colloidal-Ag	6	2	6.5
Oleate-Ag	0	0	0

3.2.2 *Minimum Inhibitory Concentration*

Minimum inhibitory concentration (MIC) is defined as the lowest concentration of an antimicrobial agent, which will inhibit the visible growth of a microorganism after overnight incubation. Determinations of minimum inhibitory concentration are considered important in diagnostic and research laboratories not only to confirm the resistance of microorganisms to an antimicrobial agent but also to monitor the activity of new antimicrobial agents. A lower MIC is an indication of a better antimicrobial agent. To determine the MIC of different forms of silver nanoparticles, the bacterium was maintained on LB agar Petridishes, fermentation was performed by inoculating a single bacterial colony from the agar plates into 5 mL liquid LB medium in a culture tube, followed by incubation at 37°C on a rotary shaker (200 rpm) for 12–18 h. A total of 200 µL of the overnight grown cells was inoculated into 10 mL of fresh LB medium. The reaction was performed by transferring 200 µL/well of the above medium into a sterile 100-well bioscreen microtitre plate along with the biogenic-Ag, colloidal-Ag, and oleate-Ag nanoparticles at varying concentrations separately. Each treatment was performed in octuplet and every experiment was repeated at least thrice to ensure reproducibility. The plate was then placed into the bioscreen plate reader and the bacterial growth was monitored every 15 min for 8 h at an optical density of 600 nm. Experiments with no silver nanoparticles served as controls. A greater lag phase and lower maximum absorbance @600 nm was observed for the colloidal-Ag nanoparticles with increase in the concentration for *E. coli* (see Table 3.1), but interestingly at the lowest concentration used, i.e., 5 µg/mL, colloidal-Ag nanoparticles killed the *B. subtilis* and biogenic-Ag nanoparticles killed both the bacterial strains, implying their higher levels of toxicity. Therefore we have performed the MIC for colloidal-Ag nanoparticles on *B. subtilis* and biogenic-Ag nanoparticles on both *E. coli* and *B. subtilis* using lower concentrations, which clearly indicated that smaller doses were sufficient enough for the colloidal-Ag nanoparticles toward *B. subtilis* and biogenic-Ag nanoparticles toward both the strains to induce toxicity effects (see Table 3.1).

As the concentration of silver nanoparticles is increased, more amount of particle is available to get adsorbed onto the surface thereby enhancing the microbial activity, however the results demonstrated that the concentration of the silver nanoparticles that prevents the bacterial growth is different for each type of bacterium, *E. coli* being more resistant than *B. subtilis* toward both colloidal-Ag as well as biogenic-Ag nanoparticles (see Table 3.1). However, there was no significant growth for *E. coli* and *B. subtilis* using colloidal-Ag nanoparticles at concentrations above 7.5 and 3.5 μg/mL respectively and biogenic-Ag nanoparticles at concentrations 4.5 and 2.5 μg/mL respectively. In another small experiment to evaluate whether the biogenic-Ag nanoparticles are toxic to the strain used to synthesize them, we have performed MIC for *S. oneidensis* as well (see Table 3.1); to our surprise the observed toxicity trends were similar to that observed for *E. coli*. We have used similar size nanoparticles with different (biogenic-Ag and oleate-Ag) and no surface coating (colloidal-Ag) to test the antibacterial activity on three different bacterial strains (*F. coli*, *S. oneidensis*, and *B. subtilis*). Interestingly, the toxicity was completely dependent on the surface coating, though all the nanoparticles used were smaller in size distribution; the toxicity varied for the type of nanoparticle used, biogenic-Ag nanoparticles turned out to be the most toxic, followed by the colloidal-Ag nanoparticles and despite being the smallest ones with average size distributions of $\sim 4 \pm 1$ nm, oleate-Ag nanoparticles did not show significant toxicity to any of the bacterial strains that were assessed. Clearly indicating that apart from size, several other factors such as the surface coat, surface charge, and surface properties might play a vital role in eliciting potential toxicity by nanoparticles toxicity that is yet to be understood thoroughly. Finally, these interesting similar silver nanoparticles mediated differential toxicity behavior, due to surface coatings, can be exploited in defining nanoparticles for a variety of intended applications, the ones that are toxic can be used as effective bactericidal agents and the ones that are not can be used in drug delivery and labeling experiments.

3.2.3 Colony Forming Units

Colony forming units (CFUs) refer to individual colonies of a microorganism (bacteria, fungi, yeast, or mold). CFUs are used as a measure of the number of colonies present in or on the surface of a sample and may be referred to as CFU per unit weight, CFU per unit area, or CFU per unit volume depending on the type of sample tested. To determine the number of CFUs, a sample is prepared and spread or poured uniformly on the surface of an agar plate followed by incubation at suitable temperature for a number of days. To evaluate the effects of any materials on the CFUs, in this case silver nanoparticles, the viability was performed in liquid cultures after treatment with various concentrations of silver nanoparticles. Aliquots were taken at different time points, for e.g., 0, 1, 5, and 24 h, and serially diluted in the appropriate growth medium, and the dilutions were plated onto LB agar Petri plates. After overnight incubation at 37°C, the number of CFUs was counted manually. We observed a toxicity trend similar to that of MIC (see Table 3.1).

3.2.4 Live/Dead Toxicity Assay

The LIVE/DEAD Bacterial Viability assay utilizes mixtures of our SYTO® 9 green-fluorescent nucleic acid stain and the red-fluorescent nucleic acid stain, propidium iodide. These stains differ both in their spectral characteristics and their ability to penetrate healthy bacterial cells. When used alone, the SYTO 9 stain generally labels all bacteria in a population; those with intact membranes and those with damaged membranes. In contrast, propidium iodide penetrates only bacteria with damaged membranes, causing a reduction in the SYTO 9 stain fluorescence when both dyes are present. Thus, with an appropriate mixture of the SYTO 9 and propidium iodide stains, bacteria with intact cell membranes stain fluorescent green, whereas bacteria with damaged membranes stain fluorescent red. The excitation/emission maxima for these dyes are about 480/500 nm for SYTO 9 stain and 490/635 nm for propidium iodide. To determine the LIVE/DEAD assay for the different silver nanoparticles, bacterial cultures were grown for 3 h in LB medium and subsequently treated with different concentrations of differently synthesized silver nanoparticles for 15 h, after which, the microbial suspension and the stain solution was added to each well of a 96-well micro plate. The plate was incubated at room temperature in the dark for 15 min, and to quantify the live and dead cells, the relative fluorescence intensities were measured using a fluorescence plate reader (excitation at 485 nm; emission at 525 and 625 nm). Similar kinds of observations were noticed when we performed the two-color Live/Dead BacLight bacterial viability assay and CFU experiments. A substantial loss of cell viability was observed for both biogenic-Ag on *E. coli* (twofold) and *B. subtilis* (threefold to fourfold) and colloidal-Ag on *E. coli* (onefold) and *B. subtilis* (twofold). Oleate-Ag did not show any significant loss of cell viability on either of the bacteria as compared to controls.

3.3 Mechanistic Investigations of Bacterium-Silver Nanoparticle Interaction

Multiple imaging techniques were performed to evaluate the molecular mechanisms that underlie the bacterial response to the silver nanoparticles.

3.3.1 Transmission Electron Microscopy Measurements

The interaction between the bacteria and the different silver nanoparticles was illustrated using bright field TEM imaging of the bacteria treated with various silver nanoparticles. Irrespective of the type of bacteria used, it was noticed that most of the nanoparticles were found attached to the surface of the bacterial cell

Fig. 3.3 Representative
TEM image showing the
interaction of *E. coli* with
biogenic silver nanoparticles.
It can be clearly seen that the
particles apparently stick to
the bacterial surfaces and
might also have internalized

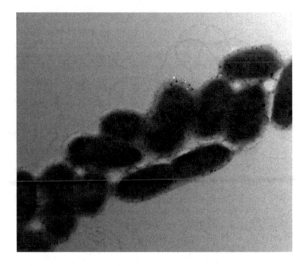

wall, implying their higher affinity toward the cells. It was obvious that the silver
nanoparticles were bound very well throughout the surface of the bacteria and were
also able to penetrate the bacteria; a similar kind of imaging technique was also
used by Morones et al. where they determined the presence of small (∼ 1 nm) silver
on the bacteria obviating the use of heavy-metal staining. Interestingly, it was also
observed that the silver nanoparticles looked well separated and spread throughout
the TEM grid prior to bacterial incubation. Upon incubation with the bacteria most
of the nanoparticles tend to agglomerate and are mostly found attached to the
surface of the bacteria. As demonstrated by electron microscopy, interaction with
silver nanoparticles resulted in perforations in the cell wall, contributing to the
enhanced antibacterial effects of the nanoparticles. Additionally, clusters of particles are seen throughout the bacterial surface. Similar observations were made for
all the different types of nanoparticles on both the bacterial strains (*E. coli and
B. subtilis*), therefore an image for the interaction of biogenic silver nanoparticles
with bacterium *E. coli* only is shown for redundancy (see Fig. 3.3).

3.3.2 Atomic Force Microscopy Measurements

Prior to atomic force microscopy (AFM) imaging the bacterial cell was immobi-
lized onto a solid mica substrate coated with gelatin, so as to aid imaging, as
described below.

3.3.2.1 Immobilization of Bacteria onto Gelatin Coated Mica

Immobilization is a commonly used technique for the physical or chemical fixation
of materials (e.g., cells, organelles, proteins, molecules) onto a solid substrate, into

a solid matrix or retained by a membrane, in order to increase their stability and make possible their repeated or continued use. Here, we immobilize bacterial cells onto gelatin coated mica so as to detect them using the AFM, for which freshly cleaved mica surfaces were dipped into 0.5% gelatin prepared in Milli Q water at 60°C and dried overnight. A total of 100 μL of the sample analyte was applied to the gelatin coated mica surface, allowed to stand for 10 min, rinsed in Milli Q water, and placed in the cell for AFM imaging.

Atomic force microscopy measurements were performed to clearly understand the depth of interaction between the bacteria and the silver nanoparticles. Several reports exist on the suitability of AFM for investigating the cell structure and morphology of both human cells as well as bacterial cells. AFM is an appropriate technique for elucidating the action of bactericides on bacterial cells. In fact, several investigations performed by various groups suggested noticeable significant changes in the cell membrane morphology upon treatment with different bactericidal agents and could be monitored with ease using AFM [1]. Bacteria with no nanoparticles served as controls where the cells looked healthy, intact with no perforations. It was observed that both *E. coli* and *B. subtilis* cells were significantly damaged with several perforations throughout the surface upon treatment (see Fig. 3.4a, b).

However, at higher magnification for both *E. coli* and *B. subtilis* cells the perforations were noticed very clearly indicating the damage caused to them by the silver nanoparticles. At higher magnification, it was also observed that though both the colloidal-Ag and biogenic-Ag nanoparticles killed the cells by causing perforations, the biogenic-Ag nanoparticles on both *E. coli* (see Fig. 3.4c) and *B. subtilis* (see Fig. 3.4d) made long scars completely tearing the membrane showing its high killing efficiency. Despite the fact that the mechanism of this interaction is still unanswered, the nanoparticles might cause perforations thereby structural changes, degradations, and finally cell death. Based on our results, though both colloidal-Ag and biogenic-Ag nanoparticles showed bacterial activity, it can be said that biogenic-Ag nanoparticles have advantageous effect than the colloidal-Ag nanoparticles, and the oleate-Ag nanoparticles were not toxic at all. Therefore, this nature of nanoparticles can be exploited for their use as efficient drugs in various drug delivery systems (biogenic-Ag) and labeling experiments (oleate-Ag).

3.4 Summary

Although we have used similar size distributions of silver nanoparticles, biogenic-Ag particles showed a profound antibacterial activity against Gram-negative (*E. coli* and *S. oneidensis* as well as Gram-positive (*B. subtilis*) bacterial strains when compared to that of chemically synthesized colloidal-Ag nanoparticles, whereas oleate-Ag nanoparticles despite being the smallest did not show significant toxicity to any of the bacterial strains that were evaluated. Our results clearly

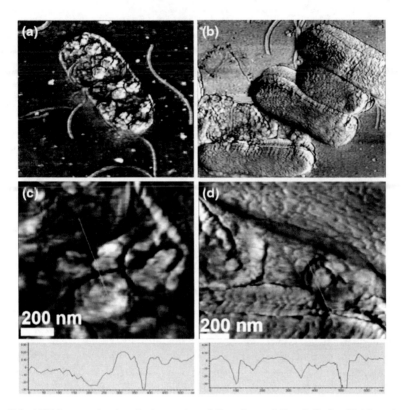

Fig. 3.4 AFM images showing the interaction of *E. coli* **a** and *B. subtilis* **b** with biogenic-Ag silver nanoparticles. **c** and **d** are the respective higher magnification images

indicate that apart from size, bioavailability, nanoparticle-cell interactions, surface coatings, and surface charge several other unknown factors might also play a prominent role in determining nanoparticles' toxicity. Although identifying novel antimicrobial agents remains a priority, the development of nanoparticles-based therapeutic systems for currently used drugs may represent cost-effective and fruitful alternatives with the major advantages being, high stability, ability to overcome multi-drug resistivity issues, etc.

Reference

1. Suresh AK, Pelletier DA, Wang W, Moon J-W, Gu B, Mortensen NP, Allison DP, Joy DC, Phelps TJ, Doktycz MJ (2010) Silver nanocrystallites: biofabrication using *Shewanella oneidensis*, and an evaluation of their comparative toxicity on Gram-negative and Gram-positive bacteria. Environ Sci Technol 44(13):5210–5215

Chapter 4
Biocompatibility and Inertness of Gold Nanocrystallites

Abstract Some types of nanoparticles are likely considered to be inert and biocompatible, such as iron oxide, ceria, and gold nanoparticles. However, the best studied and the most commonly used nanoparticles are the gold nanoparticles also known as "colloidal gold" that is being used in biomedicine for centuries. In this chapter, I will discuss about the inert nature of gold nanoparticles and the many possible explanations to justify the same. Next, our recent work on the microbial-based *S. oneidensis* mediated biosynthesis followed by physical characterizations of gold nanoparticles will be presented [1]. Finally, as an example, the inert and biocompatible nature of the biosynthesized gold nanoparticles assessed using various Gram-negative and Gram-positive bacterial cell systems will be discussed.

Keywords Bacteria · Biocompatibility · Gold nanoparticles · Inert

4.1 Inert Nature of Gold Nanocrystallites

The first and the foremost essential criteria for any nanomaterial, irrespective of its state, before it can be implemented in biomedical applications is its biocompatible or non-toxic nature. The little or no harm it does to the living cells/cellular organelles the better it is. As described earlier, nanoparticles are extensively used in various biomedical and diagnostic applications, therefore, highly inert, non-reactive, and biocompatible nanoparticles are desired, especially for the ones that advance for clinical trials. Very few nanoparticles have thus far been successful and are approved by the Federal Drug Administration (FDA) for diagnostic and clinical purposes, which includes; the magnetite nanoparticles, used for imaging, drug delivery, and photodynamic therapy; gold nanoparticles, used for drug and gene delivery, optical imaging, sensors, and photodynamic therapy; and

A. K. Suresh, *Metallic Nanocrystallites and their Interaction with Microbial Systems*, SpringerBriefs in Biometals, DOI: 10.1007/978-94-007-4231-4_4,

the ceria nanoparticles, used in bone tissue engineering. This chapter will focus and elaborate discussions on the use of gold nanoparticles as an ideal model. The biocompatible nature of gold nanoparticles has been proven by several in vitro and in vivo toxicity assessments and due to its use since the ancient times. However, the use of gold nanoparticles in modern medicine started with the invention by Robert Koch in the year 1890, a microbiologist who suggested that gold cyanide showed bactericidal activity against the bacterium, *Tubercele bacillus* that eventually led to the cure of tuberculosis in the early twentieth century. Similarly, Laude in the year 1972 used gold for the diagnosis of rheumatoid arthritis with no significant side effects. Since then gold has been used as a diagnostic agent to not only cure wide types of rheumatic diseases, such as juvenile arthritis, psoriatic arthritis, and discoid lupus erythematosus, but also used in various other medicinal applications including prosthesis in dentistry and ophthalmology, drug and gene delivery, gold coated coronary, renal, stents, etc. Toxic effects associated with gold depend on its oxidation state. Although the exact mechanism on the non-toxic nature of gold nanoparticles is unknown, one of the several deemed reasons could be its inertness; metallic gold is considered highly inert to air, water, heat, strong acids and bases (oxidation and reduction), and gold is known to be one of the least reactive metals known to man. Another reason for its stability is that metallic gold is also known to have a strong affinity for environmental rich S^- donors over O^- and N^- donors, in turn stabilization by thiolated ligands. Several reports exist on the inert and biocompatible nature of gold nanoparticles, for example, Shukla et al. suggested that the gold nanoparticles are inert or noncytotoxic to eukaryotic macrophage cells (Raw 264.7) [2]. The authors additionally suggested that gold nanoparticles reduce the production of reactive oxygen and nitrite species and do not elicit cytokine secretion, making them ideal candidates for applications in the field of nanomedicine. Likewise, Connor et al. showed that gold nanoparticles, despite being taken up by human K562 leukemia cell lines, do not cause acute cytotoxicity and are not detrimental to cellular functions [3]. Similarly, Patra et al. reported that the BHK21 (baby hamster kidney) and HepG2 (human hepatocellular liver carcinoma) cell lines were not at all affected upon treatment with gold nanoparticles [4]. In another investigation, Khan et al. while assessing the molecular effects of uptake of gold nanoparticles on HeLa cells, reported that despite being internalized by the cells, gold nanoparticles did not elicit significant cytotoxicity [5]. However, the authors suggested that gold nanoparticles might react with intracellular components and may trigger specific pathways of stress response. Although using gold nanorods, Huff et al. suggested that they were rapidly and irreversibly internalized by KB cells (oral epithelial tumor cells) [6]. The authors further stated that the internalized gold nanorods formed permanent aggregates inside the cells and were not toxic to the cells. Based on the literature it is evident that the gold nanoparticles are highly inert and impose no threat to the biosphere and lend themselves for various applications, such as biomedical, bioengineering, electronic, diagnostics, and biosensors.

4.2 Biosynthesis of Gold Nanocrystallites

Details on the biosynthesis and biosynthesis methodology, including the mechanistic of biosynthesis, were discussed earlier in Chap. 2. Tremendous interest has been generated in the use of gold nanoparticles for various biomedical and electronic applications with recent implications in optoelectronic devices, biosensors, cell labeling, drug and gene delivery, catalysis, therapeutics, and diagnostics has led increased attention to its synthesis. As discussed earlier in Chap. 1, research efforts by various investigators have demonstrated that gold nanoparticles could be synthesized in a wide range of size and shape distributions, such as rods, wires, cubes, triangles, disks, arrows, and tetrapods utilizing several physical and chemical synthesis methodologies. However, many of these synthesis techniques are cumbersome, expensive, involve the use of toxic solvents and surfactants, high temperatures and pressures, and most importantly can become unstable or aggregate upon interactions with biological materials. Therefore, an alternate approach to synthesis exploits the biosynthesis of nanomaterials and relies on natural microorganisms for the reduction of metal ions into stable nanocrystals. Biofabricated nanomaterials can not only exhibit size and shape control over a diverse array of materials but can also facilitate mass production, high yield, and reproducibility.

Herein, we describe our earlier published results on how we made use of the unique reductive phenomenon of the metal reducing bacterium, *Shewanella oneidensis,* for generating metallic gold nanoparticles [1]. *S. oneidensis* is a Gram-negative bacterium, known to have considerable potential for the bioremediation of environmental contaminants with its unique metal reducing capabilities. Due to its versatile nature it has been used for the bioreduction-based remediation of various metals, such as uranium, chromium, iodate, technetium, neptunium, plutonium, palladium, selenite, tellurite, vanadate, silver, and metal oxides of manganese and iron. Described below is the reduction of aurate (III), utilizing *S. oneidensis*, seeded with aqueous $HAuCl_4$ ions at ambient temperature and pressure to produce highly stable and biocompatible gold nanoparticles.

4.2.1 Biofabrication Process

The bacterium, *S. oneidensis* from the freezer stocks ($-80°C$) were streaked onto a Luria–Bertani agar (LBA) Petri plate and was maintained in a 30°C incubator. A single bacterial colony from an overnight LBA plate was used to inoculate 100 mL of LB broth in a 500 mL Erlenmeyer flask, followed by incubation at 30°C on a rotary shaker (200 rpm) for 24 h. The grown cells were collected by centrifugation ($5,000 \times g$, 25°C, 20 min), washed a couple of times with sterile distilled water and resuspended in 100 mL of 1 mM $HAuCl_4$ aqueous solution in a 500 mL Erlenmeyer flask and incubated at 30°C under shaking conditions (200 rpm).

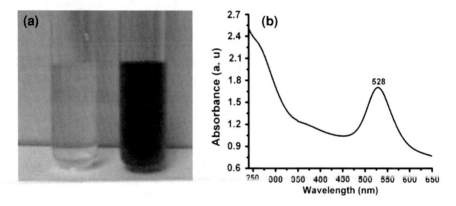

Fig. 4.1 **a** Image of the test tubes containing HAuCl$_4$ solution before (*left tube*) and after (*right tube*) the formation of gold nanoparticles by the bacterial biomass. **b** UV–Vis spectra of the biogenic gold nanoparticles

The process of biofabrication of gold nanoparticles was monitored visually and by Ultraviolet–visible (UV–Vis) absorption spectroscopy measurements were performed over a spectral sweep of 200–800 nm. After completion of the reaction process (~48 h), the reaction mixture was centrifuged (5,000×g, 20 min) to remove the bacteria, the supernatant was filtered using a sterile 0.1 μm syringe filter and the particles were collected by high speed centrifugation (100,000×g, 1 h) using an ultracentrifuge. After washing twice with Milli Q water the biogenic-Au nanoparticles were used for the described characterizations and bacterial toxicity assessment studies. To determine the percent theoretical yield of the gold nanoparticles by *S. oneidensis*, a 1 mL aliquot of the purified gold nanoparticles formed from 1 L of 1 mM HAuCl$_4$ aqueous solution was dried overnight at 60°C, followed by weighing and quantifying the mass of gold nanoparticles, relative to the amount of Au ions added.

4.3 Characterization of Gold Nanoparticles

When aqueous gold chloride solution was added to an overnight grown culture of bacterium, *S. oneidensis* the reaction mixture turned from pale yellow to deep purple (see Fig. 4.1a) over a time period of 48 h, indicating the formation of gold nanocrystallites. UV–Vis spectra of the isolated particles showed the presence of a surface plasmon resonance (SPR) exciton peak centered at 528 nm as shown in Fig. 4.1b, clearly indicating the formation of gold nanoparticles.

As discussed earlier, the synthesis mechanism might involve the bioreduction of ionic Au^{3+} to metallic Au^0 and the subsequent stabilization by capping proteins or peptides secreted by the bacteria to alleviate metal-ion mediated stress. Several related studies have demonstrated that such metal biotransformation might involve

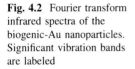

Fig. 4.2 Fourier transform infrared spectra of the biogenic-Au nanoparticles. Significant vibration bands are labeled

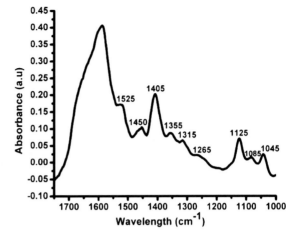

either capping proteins/peptides and reductases, quinones or cytochromes, or electron shuttles that are known to reduce various metal and metal oxides [1].

To elucidate the presence of a capping protein/peptide bound to the surface of the biogenic-Au nanoparticles, Fourier transform infrared spectroscopy (FTIR) measurements were performed. FTIR analysis revealed the presence of vibration bands centered at 1525, 1450, 1405, 1355, 1315, 1265, 1125, 1085, and 1045 cm^{-1} (see Fig. 4.2) in the region of 900–2,000 cm^{-1}. The band at 1,045 cm^{-1} corresponds to the –N–H stretch and carbonyl (–C–O–C– or –C–O–) stretch vibrations in amide linkages (amide I and amide II), a small peak for amide III was also observed at 1,355 cm^{-1}, clearly implying the presence of protein/ peptide coat on the nanoparticle surface. The protein material surrounding the nanoparticles likely serves as a capping/stabilizing agent. The bands at 1,265 cm^{-1} correspond to carbonyl and hydroxyl functional groups in alcohols and phenol derivatives.

To further determine whether the surface bound capping protein contributes to the overall stability and integrity of the nanoparticles, the gold nanoparticles were treated with a protein denaturant, 1% sodium dodecyl sulfate (SDS) detergent and boiled for 30 min at 95°C. Such a treatment resulted in the denaturation of protein leading to an immediate clumping/aggregation of the nanoparticles. It should be noted that these results not only support the presence of a protein/peptide encapping the nanoparticle surface but also demonstrates that the coating can be removed from the nanoparticle surface. The ability to remove the capping agent is necessary for applications where no surface coat is required, such as in particle annealing and thin film formation [1].

Confirmation of the formation of gold nanoparticles was based on X-ray diffraction analysis, which showed intense peak corresponding to (111), (200), and (220) in the 2θ range 20–70°, agreeing with those reported for gold nanocrystallites (see Fig. 4.3).

Fig. 4.3 X-ray diffraction of
the biogenic gold
nanoparticle powder.
Reprinted from Ref. [24] with
kind permission of © Elsevier
Ltd (2011)

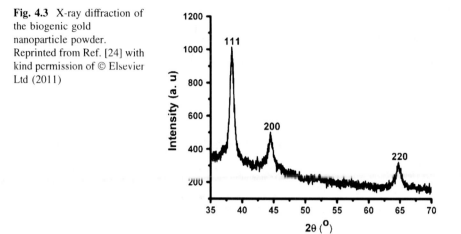

Fig. 4.4 Transmission
electron microscopy image of
the biogenic gold
nanoparticles

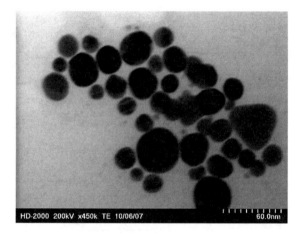

TEM micrographs of the biogenic gold nanoparticles revealed discrete, uniform, spherical nanoparticles that were well separated from each other (see Fig. 4.4). The particle size histogram distribution plot results from the counting of ∼100 particles from TEM images and revealed the particles to be in the size distribution range of ∼2–50 nm and an average of 12 ± 5 nm. The nanoparticles do not appear to be in direct contact even within the aggregates. An estimate of the size of the nanoparticles was also made from the line broadening of the (111) reflection using the Debye–Scherrer formula and is in good agreement with the nanoparticle size estimated by TEM analysis.

However, hydrodynamic sizes of the nanoparticles, as determined by DLS, appeared to be larger (∼45 nm) when compared to the TEM measurements, this can be attributed to overlapping particles and the electrical double layer phenomenon on charged particles in the DLS measurements, while TEM imaging allows latitude for eliminating aggregated particles from the analysis. The gold

Fig. 4.5 Bacterial growth curves for *E. coli* treated with different concentrations of biogenic-Au nanoparticles

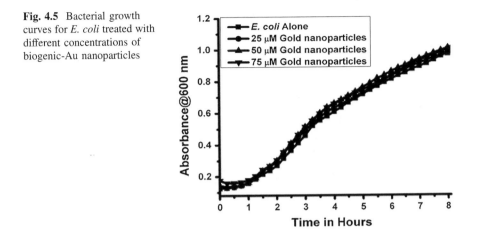

nanoparticles synthesized by *S. oneidensis* are spherical, much smaller and more homogeneous than those synthesized by other bacteria. The particles were negatively charged with a zeta potential of -16.5 ± 2 mV. Further, they are produced extracellularly and with a high yield of 88% theoretical maximum, facilitating isolation. From a one-liter culture we were able to obtain ~ 174 mg of gold nanoparticles.

Additionally, the antibacterial effects of the biosynthesized gold nanoparticles were assessed using various Gram-negative; *E. coli* and *S. oneidensis*, the strain used to synthesize them and Gram-positive; *B. subtilis* bacteria using minimum inhibitory concentrations. Based on the data interpretations no evidence of bacterial toxicity was observed for any of the three strains that were used in our studies, even at higher concentrations of 150 µM, Fig. 4.5 shows the MIC data for gold nanoparticles treated with *E. coli*.

4.4 Summary

Increasing awareness of green biofabrication processes for the synthesis of biocompatible nanomaterials emphasizes the need to develop simple and cost-effective methodologies. Biological systems have great potential for the synthesis of nanomaterials for varied applications. Although nanobiotechnology is at an early stage, described here is the demonstration of the biosynthesis of highly stable, hydrophilic, biocompatible gold nanoparticles with an average size of $\sim 12 \pm 5$ nm utilizing the metal reducing bacterium, *S. oneidensis*. This bacteria-based method of biosynthesis is straightforward, economical, and results in high yield without extreme physical conditions or detergents and solvents that are commonly associated with chemical synthesis methods. Further, bacterial toxicity assessments revealed that the biogenic-Au nanoparticles exhibited no significant bactericidal activity against *E. coli*, *S. oneidensis*, and *B. subtilis* strains. These useful features of the biofabricated gold nanoparticles and the practical synthesis methods may lend themselves to routine large-scale production and perhaps implementation in various biomedical and engineering applications.

References

1. Suresh AK, Pelletier DA, Wang W, Broich ML, Moon J-W, Gu B, Allison DP, Joy DC, Phelps TJ, Doktycz MJ (2011) Biofabrication of discrete spherical gold nanoparticles using the metal-reducing bacterium *Shewanella oneidensis*. Acta Biomater 7:2148–2152
2. Shukla R, Bansal V, Chaudhary M, Basu A, Bhonde RR, Sastry M (2005) Biocompatibility of gold nanoparticles and their endocytotic fate inside the cellular compartment: a microscopic overview. Langmuir 21(23):10644–10654
3. Connor EE, Mwamuka J, Gole A, Murphy CJ, Wyatt MD (2005) Gold nanoparticles are taken up by human cells but do not cause acute cytotoxicity. Small 1(3):325–327
4. Patra HK, Banerjee S, Chaudhuri U, Lahiri P, Dasgupta AK (2007) Cell selective response to gold nanoparticles. Nanomed Nanotechnol Biol Med 3(2):111–119
5. Khan JA, Pillai B, Das TK, Singh Y, Maiti S (2007) Molecular effects of uptake of gold nanoparticles in HeLa cells. Chembiochem 8(11):1237–1240
6. Huff TB, Hansen MN, Zhao Y, Cheng J-X, Wei A (2007) Controlling the cellular uptake of gold nanorods. Langmuir 23(4):1596–1599

Chapter 5
Engineered Metal Oxide Nanocrystallites: Antibacterial Activity and Stress Mechanism

Abstract Nanoparticles are being developed for several research as well as medicinal and engineering implications. Following them a host of new potential health issues due to their size-dependent larger surface area and high reactivity. In this brief chapter, an introduction on the likely interactions of nanoparticles with biotic environment, various possibilities of these man-made nanoparticles coming in contact with the environment and thereby consequences will be discussed. This will be illustrated using our recent work on the effects of various sizes of engineered cerium oxide nanoparticles on the growth and viability of several Gram-negative and Gram-positive microorganisms. The relation between the growth inhibition, reactive oxygen species (ROS) generation, and up- and/or downregulation of transcriptional stress genome using *E. coli* will be discussed. How different cerium oxide nanocrystallites were synthesized by solvent-free hydrothermal-based approach so as to eliminate cross contamination from the use of toxic solvents and surfactants. Further, utilization of advanced technique like the transmission electron microscopy and microarray-based transcriptional profiling to evaluate the bacterial response mechanisms will be described.

Keywords Antibacterial · Biotic environment · Metal oxide · Stress mechanism

5.1 Engineered Nanoparticles and the Environment

Engineered products of nanomaterials are beginning to pervade several aspects of life and technology. Nanomaterials are becoming integral to technologies such as biomedical engineering, electronics, biosensors, and as common household consumer products such as sun screens, paints, ointments and creams, and stain-resistant clothing. Among the broad variety of nanomaterials, nanoparticles represent one of

A. K. Suresh, *Metallic Nanocrystallites and their Interaction with Microbial Systems*,
SpringerBriefs in Biometals, DOI: 10.1007/978-94-007-4231-4_5,
© The Author(s) 2012

the largest and the most widely used classes. Metal and semi-conductor metal oxide or metal sulfide nanoparticles with novel properties enable their far reaching applications such as in optoelectronic, catalysis, and solar energy conversion. Similarly, biomedical applications of nanoparticles abound and include cancer therapies, image contrast agents, and drug and gene delivery. The reactivity, cohesive energy, and physical properties of nanoparticles can be significantly altered. Nanoparticles-based catalysts are used broadly for producing chemicals, pollution control, and energy conversion. Along with industrial and sewage wastes, the use of nanoparticles is ubiquitous and found in consumer products for removing bathroom odors, in swimming pools, water remediation, and for reducing toxic emissions from automobile exhausts. For example, every automobile produced since the early 1970s contains a catalytic converter that oxidizes hydrocarbons and carbon monoxide and reduces nitrous oxide from exhaust systems. Further, catalysis is seen as the foundation pillar of green chemistry that seeks to produce materials in a manner that does not compromise human health or the bioenvironmental threat. Despite the critical role played by metal nanoparticles in catalysis, relatively little is known regarding the fate and transformation of these engineered structures in the environment. Most likely, metal nanoparticles might also interact with microbial systems, and therefore are intimately going to be associated with the bacteria. Such interactions will ultimately influence the mobility and fate of metals in the environment. Some properties that make nanoparticles useful in a variety of applications, however, can potentially make them harmful to the biosphere. Therefore, researchers have also attempted to evaluate the risks associated with these particles, before they can be applied, so as to prevent environmental damage and health hazards. Tons of literature have already been published on the toxicological impacts of nanomaterials along with reviews and perspectives. A huge number of studies has addressed the potential toxicity of nanoparticles on different cell systems including bacteria, fungi, mammalian cells, ecosystems, and also using whole organisms. On the other hand, when it comes to a whole organism or animal it is hard to test nanomaterials toxicity effects, as there is so much of synthetic engineering involved and varies from individual research groups and it is also hard to get the proprietary information on the synthesis strategy. However, with all these particles and different cell systems, the potential mechanism of toxicity has been attributed to several possible mechanisms; the dissolution or release of ions from the nanoparticles elicit either inflammatory response, lipid peroxidation, mitochondrial dysfunction, disruption of cell-membrane integrity, oxidative stress, protein or DNA binding and damage, reactive oxygen species (ROS) generation, finally apoptotic cell death. It seems likely that the mechanism of toxicity depends on properties of the nanoparticles, such as the surface area, size and shape, capping agent, surface charge, purity of the particles, structural distortion, and bioavailability of the individual particles. Despite efforts by several investigators, many open questions and controversies in relation to nanoparticles toxicity still remain and the exact mechanisms and material dependence of nanoparticle-induced toxicity remain unclear.

The proliferation of nanotechnology has prompted researchers over the safety of these engineered nanomaterials to both mankind as well as the environment. The rate at which research in nanoscience and nanotechnology is progressing and being utilized for some or the other applications, it is almost inevitable that living beings will be exposed to these nanomaterials. The same properties that make these nanoparticles useful in various applications can potentially have adverse effects on the environment. Nevertheless, a better understanding on the hazards associated with individual nanomaterial type might reduce damage caused to the biosphere or toxic effects on health.

This chapter will describe in detail using an example of our work the effects of different size distributions of engineered cerium oxide nanoparticles on the growth and viability of various Gram-negative (*Escherichia coli, B. subtilis*) and Gram-positive (*S. oneidensis*) bacterial strains. Cerium oxide (CeO_2) nanoparticles are a prime example of a metal oxide nanomaterial that is being developed for a wide range of novel industrial and biomedical uses due to its large surface area, and semiconductor nature, which in turn lead to several interesting properties. It is used extensively as an abrasive in semiconductor manufacturing, as a catalyst for automobile exhaust, as fuel additive to induce combustion, as a UV-light absorber, and as fuel cell electrolytes. Anti-oxidative properties of cerium oxide nanoparticles have been shown to have biomedical applications, such as in protecting cells against radiation damage, oxidative stress, and inflammation [1].

The present chapter will also focus on the antibacterial assessments of cerium oxide nanoparticles which we illustrate by the several aims that were used to evaluate different appropriate approaches for assessing bacterial toxicity using well-characterized nanoparticles and multiple standard bacterial assay systems. Specifically, how the presence of cerium oxide nanoparticles affects the growth and viability of several microbial strains by examining bacterial growth in the presence and or absence of nanoparticles will be described. Additionally, studies such as the stress responsive generation of ROS, transmission electron microscopy (TEM) imaging of the cells, and microarray-based transcriptional profiling that potentially reveal in detail the mechanism of their interaction and the genetic response of the bacteria to cerium oxide nanoparticle stress are ascribed.

5.2 Synthesis and Characterization of Cerium Oxide Nanoparticles

Cerium oxide nanoparticles were synthesized through a modified surfactant and template-free synthesis methodology [1]. Briefly, cerium hydroxide $Ce(OH)_3$ was prepared by mixing 25 mL of $NH_3 \cdot H_2O$ or NaOH solution with 25 mL of $Ce(NO_3)_3$ solution under vigorous stirring (5.1). After 30 min of equilibration, $Ce(OH)_3$ precipitates were separated by centrifugation, washed a couple of times with Milli Q water, and resuspended in 50 mL of Milli Q water while sonicating.

To convert $Ce(OH)_3$ to CeO_2, the cerium hydroxide colloidal suspension was heated in Teflon-lined containers (5.2).

$$Ce^{3+} + OH^- \rightarrow Ce(OH)_3 \dots\dots\dots\dots (\text{precipitation}) \qquad (5.1)$$

$$Ce(OH)_3 \rightarrow CeO_2 + H_2O \dots\dots\dots\dots (\text{hydrothermal dehydration}) \qquad (5.2)$$

Upon completion of the reaction process, the CeO_2 nanocrystals were dialyzed against Milli Q water to remove excess free ions (until electric conductivity <2 μS) using a 10,000 molecular weight cut off membrane. Batches of nanoparticles with nominal sizes of approximately 6 ± 3.5, 15 ± 4.3, $22 + 5.7$, and 40 ± 10 nm (as measured by electron microscopy) were prepared. The smallest and the next smallest CeO_2 nanocrystals were made by using 0.05 and 0.5 M of $Ce(NO_3)_3$, respectively, followed by slow addition of 29.5% ammonia solution during the precipitation process, and finally by hydrothermal treatment at a relatively low temperature 140°C for short time (2 h). Whereas, the next two larger particles of CeO_2 nanocrystals were made by using 0.05 M and 0.5 M of $Ce(NO_3)_3$, respectively, through a rapid precipitation process using 2.5 M NaOH, followed by hydrothermal treatment at relatively high temperatures 200°C and longer times (80 h). The synthesized CeO_2 nanoparticles were further purified by centrifugation, washing couple of times with Milli Q, and dialysis against Milli Q to remove any soluble impurities and used further.

The synthesized cerium oxide nanoparticles were characterized in terms of purity, morphology, dimensions, and surface charge using various characterization and analytical techniques described earlier in Chap. 1. To determine the size and shape distribution of the cerium oxide nanoparticles drop-coated particles suspension on carbon-coated copper grids were imaged using the transmission electron micrograph. TEM images (see Fig. 5.1) of the four sizes of nanoparticles revealed differences in the structure of the polydispersed nanoparticles. The smallest particles with average size distribution, 6 ± 3.5 nm, were more or less square shaped, with a few ovoid shapes observed in the TEM images (see Fig. 5.1a). The next largest particle showed a preponderance of circular and ovoid shapes with an average of 15 ± 4.3 nm. Additionally, edges are often apparent on the ovoid particles and a few cuboidal particles were evident (see Fig. 5.1b). The next largest set of particles showed ovoid particles as well as additional shapes including rectangular and triangular particles with size distributions of 22 ± 5.7 nm (see Fig. 5.1c). The largest particles continued the trend and revealed diverse particle shapes (see Fig. 5.1d) and the average sizes of 40 ± 10 nm; further, TEM images showed crystalline particles.

To further elucidate the purity and surface properties of the cerium oxide nanoparticles energy dispersive X-ray spectroscopy was performed, which revealed the presence of prominent peaks for Ce and O (see Fig. 5.2), clearly revealing the presence of elemental cerium oxide nanoparticles. As mentioned above, our methodology is a modified template or capping agent-free synthesis. The resulted cerium oxide nanoparticles are pure and do not involve any external

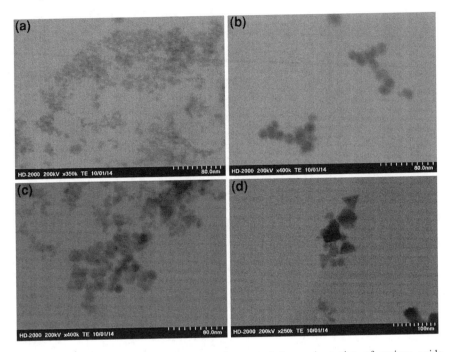

Fig. 5.1 Transmission electron microscopy images of the various size of cerium oxide nanoparticles. **a** 6 ± 3.5 nm, **b** 15 ± 4.3 nm, **c** 22 ± 5.7 nm and **d** 40 ± 10 nm

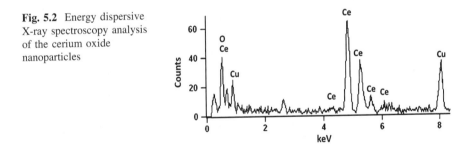

Fig. 5.2 Energy dispersive X-ray spectroscopy analysis of the cerium oxide nanoparticles

stabilizing agent encapped on their surface. This can also be observed based on energy dispersive X-ray spectroscopy analysis (see Fig. 5.2). Additional peaks for elemental Cu were also seen due to copper on which the samples were prepared.

The charge of the second largest cerium oxide naoparticles as measured by the zeta potential measurements was determined to be $\sim +39 \pm 4.5$ mV implying that they have overall net positive charge. Interactions of nanoparticles with foreign materials, primarily metal oxide nanoparticles is well documented due to their ability to absorb foreign materials, and are known to drastically influence the surface properties of nanoparticles. Therefore, the impact of standard protein, Bovine Serum Albumin (BSA), on the overall charge and hydrodynamic size

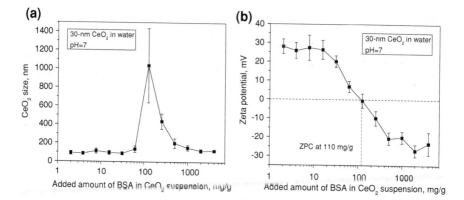

Fig. 5.3 Influence of Bovine Serum Albumin on the size and charge of the cerium oxide nanoparticles. **a** Dynamic light scattering measurement to determine size distributions and **b** Zeta potential measurements to measure the overall surface charge

distributions of the particles are assessed following dynamic light scattering and zeta potential measurements. With the increase in concentration of BSA, the nanoparticles showed increased aggregation behavior as revealed by their increase in hydrodynamic size distributions (see Fig. 5.3a). Similarly addition of BSA had influence on the overall charge of the nanoparticles, where the charge of the particles shifted from net positive (+39 ± 4.5 mV) to net-negative charge (+26 ± 5 mV) (see Fig. 5.3b).

5.3 Bacteriological Toxicity Assessment

Details on the description of the various toxicity evaluation methods along with the principle mechanism are described earlier. Therefore, only the bactericidal toxicity assessments performed using the cerium oxide nanoparticles are described below.

5.3.1 Disk Diffusion and Minimum Inhibitory Concentration Assays

The antibacterial activity on the Gram-negative (*E. coli, S. oneidensis*) and Gram-positive (*B. subtilis*) bacteria was compared for variously sized cerium oxide nanoparticles using the diameter of zone of inhibition (DIZ) in disk diffusion assay and minimum inhibitory concentration assays. The strains susceptible to disinfectants showed larger DIZ (see Fig. 5.4a), whereas resistant strains exhibit smaller or no DIZ. However, in studies where growth inhibition was observed, the

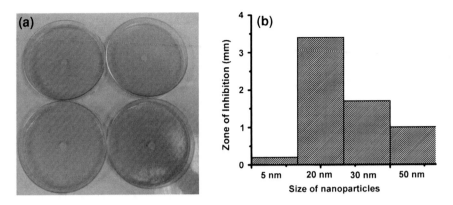

Fig. 5.4 a Image of an agar plate containing the different sized cerium oxide nanoparticles impregnated disks showing the diameter of the inhibition zone. **b** Quantitative data analysis on the zone of inhibition shown by the different cerium oxide nanoparticles on *E. coli*

bacterial inhibition was concentration dependent, with concentrations ranging from 50 to 150 mg/L. For *E. coli*, the largest zone of inhibition was observed with disks impregnated with 15 ± 4.3 nm cerium oxide nanoparticles (see Fig. 5.4). The next largest zone of inhibition was found for the 22 ± 5.7 nm particles, followed by the 40 ± 10 nm size particles while the smallest 6 ± 3.5 nm particles showed no inhibition, whereas for *B. subtilis*, the trend was completely opposite, bacteria treated with smallest and the largest nanoparticle samples showed maximum inhibition, followed by the 22 ± 5.7 nm sized particles, and 15 ± 4.3 nm particles were not at all inhibitory by this assay. For *S. oneidensis*, none of the nanoparticle samples showed growth inhibition in the DIZ assay [1].

Minimum inhibitory concentration assays were performed in test tube cultures. The four sizes of nanoparticles, at varying nanoparticle concentrations (50, 100, 150 mg/L), were added to log phase bacterial cultures, and the bacterial growth was monitored at wavelength of 600 nm. As cerium oxide nanoparticles particles tend to absorb foreign materials, the assay for these particular nanoparticles were performed using two different growth medium, the regular complex LB medium and the minimal medium, separately. The MIC assay with all the different types of nanoparticles and using any bacterial strain exhibited no inhibition for any concentration used in LB medium that are in agreement with the literature [1]. Thill et al. reported that the interaction of the nanoparticles with the organic material components in the LB medium is responsible for their non-bactericidal activity against the bacterium *E. coli* [1]. In minimal media the only organic carbon source is glucose. For all the three bacteria that were evaluated, the observed growth inhibition trends were similar to that observed by the DIZ assay.

Overall, the physical characterizations of the different CeO_2 nanoparticles indicate the heterogeneities and chemical properties associated with the samples that may need to be accounted for when interpreting toxicity data. Clearly, even with refined synthesis procedures, a range of physical structures is present.

These structures may have different biological reactivity that can complicate interpretations. Further, the point of zero charge of the particles is near neutral pH values
and can cause the particles to agglomerate at pH values optimal for bacterial growth.
This point of zero charge can also shift in different media used for bacterial growth
indicating the presence of other media components competing for binding sites on the
particles.

5.4 Mechanism of Toxicity

To begin to evaluate the molecular mechanisms that underlie bacterial response to
the CeO_2 nanoparticles, series of imaging experiments and molecular analyses was
performed. These assessments focused on the effects of the 15 ± 4.3 nm sized
CeO_2 nanoparticles on E. coli due to its demonstrated inhibition on cell growth.
Although such analyses are often involved, an analysis of the molecular mechanisms can ultimately be used to classify bacterial response mechanisms.

5.4.1 Reactive Oxygen Species Generation

ROS are chemically reactive molecules such as peroxides that contain oxygen. ROS
are highly reactive due to the presence of unpaired valence shell electrons. ROS form
as a natural by-product of the normal metabolism of oxygen and have important roles
in cell signaling, homeostasis, and also apoptosis. However, during times of environmental stress, in the present case in the form of nanoparticles, ROS levels are
known to increase drastically which might result in significant damage to cell
structures. This cumulates into a situation known as oxidative stress. ROS production
can be monitored using various analytical techniques. In our study the ROS
production upon exposure of the bacterial suspension to smallest cerium oxide
nanoparticles was monitored by change in the color of 2,3-bis(2-methoxy-4-nitro-5-
sulfophenyl)-2H-tetrazolium-5-carboxanilide (XTT) due to reduction of superoxide
(O^{2-}) to XTT-formazan. Briefly, in a 96-well plate, samples containing various
concentrations of the cerium oxide nanoparticles, 100 μM XTT in 200 μL of
appropriate medium, were monitored for change in the absorbance at 470 nm using a
spectrophotometer at various time intervals, which is indicative of superoxide
production.

The generation of ROS upon bacterial interaction with 15 ± 4.3 nm sized
cerium oxide nanoparticles when examined using an XTT assay yields a colorimetric signal when reduced by superoxides. Using this assay, involving E. coli, we
found no signal for oxidative stress by generating superoxide. ROS are best known
to implicate toxicity to several prokaryotic and eukaryotic systems upon interaction with metal/metal oxide nanoparticles. ROS in either the form of superoxide
radical (O^{2-}), hydrogen peroxide (H_2O_2), and hydroxyl radical $(OH\cdot)$ causes

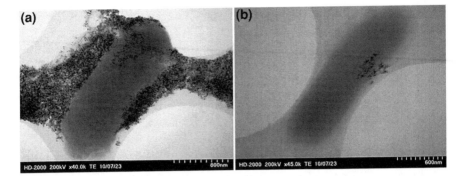

Fig. 5.5 Transmission electron microscopy images of *E. coli* upon exposure to 15 ± 4.5 nm cerium oxide nanoparticles to assess nanoparticles-microbe interaction. Clearly, positive interaction of nanoparticles with cells was observed

oxidative stress, thereby causing damage to DNA, cell membranes, cellular proteins, and finally leading to cell death. The presence of ROS was examined using an XTT assay, which yields a colorimetric signal when reduced by superoxides.

5.4.2 Mode of Interaction Based on Transmission Electron Microscopy

To assess if there is a potential mode of interaction between the CeO_2 nanoparticles and *E. coli*, transmission electron microscopy imaging experiments of the nanoparticle-treated bacteria were performed. Effective imaging required refinement of the experimental protocol. Initial experiments involved placing a droplet of bacteria/nanoparticles on the carbon-coated grids and air-drying. However, this resulted in either a film or a precipitate from the media obscuring nanoparticle interactions with the bacteria. To eliminate this problem, the nanoparticle/bacterial culture was pelleted and resuspended in water prior to placing on the grid. However, images of these samples showed that nanoparticles in the solution migrate toward the edges of the bacteria forming a halo around the perimeter of the bacteria, presumably during drying of the sample. By allowing a droplet of the bacteria containing solution to first settle on the copper or nickel carbon-coated grid and then rinsing the surface by plunging once into water eliminated this artifact and removed unbound nanoparticles and bacteria.

Figure 5.5 shows the representative TEM image of *E. coli* grown in M9 treated with 15 ± 4.3 nm sized cerium oxide nanoparticles using this procedure. The TEM image suggests that the nanoparticles adsorbed to, but do not penetrate, the bacterial cells. Additionally, clusters of particles are seen, consistent with the light scattering-based characterizations (see Fig. 5.5a). The imaging results obtained are consistent with those presented by Thill and colleagues, who additionally suggest that the adsorption of nanoparticles to the bacterial cell walls accounts for their toxicity [1]. Although STEM imaging provides a direct measure of nanoparticle interactions with the bacteria, the potential for imaging artifacts cannot be eliminated.

5.4.3 Transcriptional Microarray Analysis

Microarray is a multiplex technology used in molecular biology and in medicine. It consists of an arrayed series of thousands of microscopic spots of DNA oligo-nucleotides, called features, each containing picomoles of a specific DNA sequence. This can be a short section of a gene or other DNA element that is used as probes to hybridize a cDNA or cRNA sample (called target) under high stringency conditions. Probe-target hybridization is usually detected and quantified by detection of either fluorophore or silver or chemiluminescence-labeled targets to determine relative abundance of nucleic acid sequences in the target. In standard microarrays, the probes are attached to solid surface by a covalent linking to a chemical matrix using cross-linkers, such as epoxy-silane, amino-silane, lysine, polyacrylamide, etc. The solid surface can be glass or a silicon chip, commonly known as gene chips. Microarrays can be used to measure changes in expression levels, to detect single nucleotide polymorphisms, in genotyping or in rese-quencing mutant genomes. DNA microarrays can be used to detect DNA (as in comparative genomic hybridization), or detect RNA (most commonly as cDNA after reverse transcription) that may or may not be translated into proteins. The process of measuring gene expression via cDNA is called expression analysis or expression profiling. For discovery of genetic-based response mechanisms, the global transcriptomic response of E. coli upon exposure to cerium oxide nano-particles was assessed using whole-genome microarray analysis and compared to treatments with cerium chloride or Milli Q water.

5.4.3.1 Microarray Hybridization Methodology

For microarray experiments an overnight grown culture of E. coli was used to inoculate into 100 mL of prewarmed LB medium to an optical density of ~ 0.096 (at 600 nm) and incubated at 37°C on a shaker at 200 rpm until mid-log phase (at 600 nm ~ 0.5). Cultures were treated separately with either pre-warmed cerium oxide nanoparticles or cerium chloride suspension at a little higher concentration than required to induce minimum inhibition, ~ 100 mg/L, and Milli Q water alone. After an hour of treatment, cells were collected by centrifugation (5000 g, 2 min at 4°C) and snap-freezing using liquid nitrogen. Three separate controls and three experimental cultures were determined for every condition. Total cellular RNA was isolated by incubating the cells with 1 mg/mL of lysozyme to lyse the cells. Purified, fluorescently labeled cDNA was hybridized to

E. coli K12 gene expression 4×72 K arrays using a Nimblegen hybridization system, according to the manufacturer's instructions. Microarrays were washed using buffers of increasing stringency, microarray mixes were removed at 42°C in Nimbelgen Wash Buffer I, washed manually in room temperature buffers; Wash Buffer I for 2 min, Wash Buffer II for 1 min, Wash Buffer III for 15 s, dried for 80 s using a Maui Wash System, scanned with a Surescan high-resolution DNA microarray scanner, and the images were quantified using the Nimblescan software. Raw microarray data was $\log_2$ transformed and imported into the statistical analysis software JMP Genomics 4.0. Microarray data were normalized using the Lowess normalization algorithm within JMP Genomics and an analysis of variance (ANOVA) was performed to determine significant differences in gene expression levels between conditions and time points using the FDR testing method ($p < 0.01$).

5.4.3.2 Stress Genomic Analysis

For the identification of genomics based toxicity response mechanism, the global transcriptiomics of *E. coli* treated 15 ± 4.3 nm sized cerium oxide nanoparticles was evaluated based on whole genomic microarray analysis and compared to similar treatments with cerium chloride or water. In the microarray experiments, there was only a slight impact on cell growth and no appreciable differences in culture responses to the respective treatments. The entire microarray data set has been deposited in the Gene Expression Omnibus (GEO, http://www.ncbi.nlm.nih.gov/geo/) database. Overall, 144 differentially expressed genes at statistically significant values ($-\log 10(p) > 3.8$) in an analysis of variance (ANOVA) model using a stringent false discovery rate testing method ($\alpha = 0.01$) were identified. Of these genes, 62 showed twofold or greater differences in relative gene expression for all the pairwise comparisons.

Analysis of the microarray data indicates that cells that received either the CeO_2 nanoparticle or the $CeCl_3$ treatments had higher levels of cydA and cydB transcripts than cells treated with water (see Table 5.1). cydA and cydB expressions are known to be induced by iron limitation and oxidative stress exposure [1]. The increased abundance in transcripts encoding NirD and the high-affinity terminal oxidase cytochrome bd-I used under microoxic conditions, with the concomitant decrease expression for succinate dehydrogenase and cytochrome b terminal oxidase genes is indicative of the cerium (either in nanoparticle or ionic form) disrupting *E. coli* respiration, iron limitation, or oxidative stress (see Table 5.1). Consistent with the interpretation of the cerium salt exposure data, the presence of cerium, in the form of a nanoparticle or ion, interacts with *E. coli* and alters electron flow and respiration (see Table 5.1). The groES and sodA genes showed greater expression levels following treatment with water compared to either of the other two treatments, which is indicative that the latter two did not elicit major oxidative stress responses under the conditions used in this study. This observation

Table 5.1 Pair wise comparison of differential gene expression for selected genes; Control/CeCl$_3$, CeO$_2$/CeCl$_3$ and Control/CeO$_2$

Gene	Product	Cont/CeCl$_3$	CeO$_2$/CeCl$_3$	Cont/CeO$_2$
cyoA	Cytochrome ubiquinol oxidase, subunit II	0.5	1.0	0.5
sdhD	Succinate dehydrogenase, cytochrome small membrane subunit	0.4	1.5	0.4
sdhC	Succinate dehydrogenase, cytochrome large membrane subunit	0.3	1.4	0.5
sdhA	Succinate dehydrogenase, flavoprotein subunit	0.4	1.3	0.5
sdhB	Succinate dehydrogenase, fes subunit	0.4	1.2	0.5
cydA	Cytochrome d terminal oxidase, subunit I	2.0	1.0	1.9
nirD	Nitrate reductase	1.6	1.1	1.8
cydB	Cytochrome d terminal oxidase, subunit II	2.3	0.9	2.0

Reprinted from Ref. [69] with kind permission © The American Society for Microbiology (2010)

Fig. 5.6 Venn diagram of significant genes expressed upon treatment of 15 ± 4.3 nm cerium oxide nanoparticles with *E. coli*

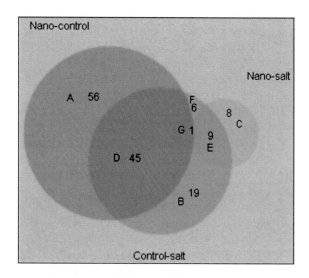

is in agreement with the XTT assays, which showed no detectable reactive oxygen species.

The greatest genetic-based response came from the water treatment, which represents an osmotic shock treatment to *E. coli* growing in M9 medium (see Fig. 5.6). A proportional Venn diagram analysis showed that many of the iron uptake genes responded significantly in the nanoparticle treatment compared to the water treatment (see Fig. 5.6), while the cus operon responded significantly in both the CeO$_2$ nanoparticle and salt treatments compared to the control treatment. There was no significant differential gene expression from these loci when the CeO$_2$ nanoparticle and salt treatments were compared. Eight genes (rnt, thiS, cysW, yciW, cysI, ilvG, cysN, pyrB) were significantly differentially expressed between the CeO$_2$ nanoparticle and salt treatments. The majority of these genes are involved in sulfur metabolism and as other related genes such as cysD and cysJ responded to other treatments. A number of previous studies have shown relationships between genes involved in sulfur metabolic processes, iron uptake, respiration and different stress responses [1].

5.5 Summary

Tremendous use of nanoproducts emphasizes the need to understand the interactions between these materials and living systems and the need to standardize methods for toxicity assessments. The present investigation examined the potential toxicity of engineered cerium oxide nanoparticles in relation to their size on different bacteria. The importance of using well-characterized nanomaterials of a known synthesis route is highlighted. Materials of commercial origin may contain unknown surfactants or additives that can obscure toxicity interpretations. Hydrothermally prepared cerium oxide nanoparticles showed growth inhibition toward *E. coli* and *B. subtilis* in minimal media and not LB medium, as a function of the nanoparticle size. The observed size-dependent response may result from size-dependent characteristics of the cerium oxide nanoparticles and/or metabolic characteristics of the different organisms. However, the role of the different media used to grow these bacteria and the interactions of the nanoparticles with these media must be considered when assessing bacterial response. In contrast, *S. oneidensis* growth was not inhibited by the cerium oxide nanoparticles. Our observations suggest that nanoparticle interactions with bacteria are strain dependent. Redundant measures of bacterial growth and toxicity in the presence of the different materials support this observation. Further investigation into the mechanism of growth inhibition for *E. coli* showed that nanoparticle-bacterial interactions likely occur and that a general stress response is elicited. Extending this general approach of using well-characterized materials, multiple organisms and measures of growth and viability to other nanomaterials will be critical for understanding the interaction of nanomaterials with living systems and for interpreting the effect and eventual fate of engineered materials in the environment.

Reference

1. Pelletier DA, Suresh AK, Holton GA, McKeown CK, Wang W, Gu B, Mortensen NP, Allison DP, Joy DC, Allison MR, Brown SD,Phelps TJ, Doktycz MJ (2010) Engineered cerium oxide nanoparticles: effects on bacterial growth and viability. Appl Environ Microbiol 76(24): 7981–7989

Printed by Publishers' Graphics LLC USA
MO20120404-678
2012